ISSN 0044-2267

ZEITSCHRIFT FÜR ANGEWANDTE MATHEMATIK UND MECHANIK

INGENIEURWISSENSCHAFTLICHE FORSCHUNGSARBEITEN

BAND 60 1980 HEFT 5

AUS DEM INHALT

H A U P T A U F S Ä T Z E

A K A D E M I E - V E R L A G · B E R L I N

ZAMM Bd. 60 Nr. 5 S. 221—276 Berlin, Mai 1980

EVP 18,— M 34115

INHALT

Hauptaufsätze Seite

Wir bitten, **Manuskriptsendungen** zweifach (Original und eine Kopie, sprachlich einwandfrei, Formeln mit Maschine oder in Druckschrift geschrieben) an folgende Anschrift zu richten:

Zeitschrift für Angewandte Mathematik und Mechanik z. Hd. Herrn Prof. Dr. Günter Schmidt
Zentralinstitut für Mathematik und Mechanik an der Akademie der Wissenschaften der DDR
DDR-1080 Berlin, Mohrenstraße 39.

Zu den Arbeiten, die als Hauptaufsätze bestimmt sind, ist auf gesondertem Blatt eine Zusammenfassung von 5 bis 10 Zeilen in englischer und (möglichst) deutscher und russischer Sprache beizufügen. Ausführliche Hinweise für die Autoren, um deren strikte Berücksichtigung gebeten wird, finden sich im Anschluß an das Inhaltsverzeichnis des Jahrganges **59** (1979).

Die Autoren erhalten von den Hauptaufsätzen 75, von den Kleinen Mitteilungen 25 Sonderdrucke ohne Berechnung, darüber hinaus weitere Sonderdrucke gegen Berechnung.

Der Verlag behält sich für alle Beiträge das Recht der Vervielfältigung, Vorbereitung und Übersetzung vor.

Bestellungen sind zu richten

- **in der DDR** an den Postzeitungsvertrieb; an eine Buchhandlung oder an den AKADEMIE-VERLAG, DDR-1080 Berlin, Leipziger Str. 3—4
- **im sozialistischen Ausland** an eine Buchhandlung für fremdsprachige Literatur oder an den zuständigen Postzeitungsvertrieb
- **in der BRD und Westberlin** an eine Buchhandlung oder an die Auslieferungsstelle KUNST UND WISSEN, Erich Bieber, 7000 Stuttgart 1, Wilhelmstraße 4—6
- **in Österreich** an den Globus-Buchvertrieb, 1201 Wien, Höchstädtplatz 3
- **in den übrigen westeuropäischen Ländern** an eine Buchhandlung oder an die Auslieferungsstelle KUNST UND WISSEN, Erich Bieber GmbH, Ch-8008 Zürich/Schweiz, Dufourstraße 51
- **im übrigen Ausland** an den Internationalen Buch- und Zeitschriftenhandel; den Buchexport, Volkseigener Außenhandelsbetrieb der Deutschen Demokratischen Republik, DDR-7010 Leipzig, Postfach 160, oder an den AKADEMIE-VERLAG, DDR-1080 Berlin, Leipziger Straße 3—4

ZEITSCHRIFT FÜR ANGEWANDTE MATHEMATIK UND MECHANIK

Herausgeber und Chefredakteur: Prof. Dr. Günter Schmidt. Redaktion: Dr. Winfried Heinrich, Dr. Horst Weinert, Dipl.-Math. Friedhild Dudel, Helga Rühl, Zentralinstitut für Mathematik und Mechanik der Akademie der Wissenschaften der DDR. Verlag: Akademie-Verlag, DDR-1080 Berlin, Leipziger Straße 3—4; Fernruf: 2236221 oder 2236229 Telex-Nr.: 114420; Bank: Staatsbank der DDR, Berlin, Kto.-Nr.: 6836-26-20712. Anschrift der Redaktion: Zentralinstitut für Mathematik und Mechanik der Akademie der Wissenschaften, DDR-1080 Berlin, Mohrenstraße 39; Fernruf: 2000561.

Veröffentlicht unter der Lizenznummer 1282 des Presseamtes beim Vorsitzenden des Ministerrates der Deutschen Demokratischen Republik. Gesamtherstellung: VEB Druckerei „Thomas Müntzer", 5820 Bad Langensalza. Erscheinungweise: Die Zeitschrift für Angewandte Mathematik und Mechanik erscheint monatlich. Die 12 Hefte eines Jahres einschließlich Tagungsheft bilden einen Band. Bezugspreis je Band 360,— M zuzüglich Versandspesen (Preis für die DDR 216,— M). Bezugspreis je Heft 30,— M (Preis für die DDR 18,— M). Bestellnummer dieses Heftes: 1009/60/5.

AN (EDV) 35937

ASMW

ISSN 0044-2267

ZAMM 60, 221–234 (1980)

H. F. BAUER / ST. METTEN / J. SIEKMANN

Dynamic Behavior of Distensible Fluid Lines Carrying a Pulsating Incompressible Liquid *)

An analysis of the propagation of pressure waves through a thin-walled elastic tube, filled with a streaming incompressible fluid, is presented. The outer surface of the tube is assumed to be free. Axially symmetric solutions for the linearized equations governing the motion of the (viscous or nonviscous) fluid and the (elastic) solid are obtained. The motion of the tube wall is described by the DONNELL shell equations. The solutions lead to a complicated transcendental dispersion equation relating wave number and frequency. This equation is studied by both analytical and numerical means in some detail.

Vorliegende Arbeit befaßt sich mit der Ausbreitung von Druckwellen in einem dünnen elastischen Rohr, das mit einer strömenden, dichtebeständigen Flüssigkeit gefüllt ist. Die Außenwand des Rohres soll als spannungsfrei angenommen werden. Die axialsymmetrischen Lösungen der linearisierten Grundgleichungen für Flüssigkeit und Festkörper werden angegeben, wobei die Flüssigkeit als reibungsbehaftet bzw. reibungsfrei vorausgesetzt wird. Die Bewegung der Schalenwand werde durch die DONNELLschen Schalengleichungen beschrieben. Die Lösungen führen auf die sogenannte Dispersionsgleichung, welche Wellenzahl und Frequenz miteinander in Beziehung setzt. Resultate analytischer und numerischer Untersuchungen der Dispersionsgleichung werden mitgeteilt.

В настоящей работе рассматривается распространение ударных волн в тонкой упругой трубе, через которую протекает жидкость с постоянной плотностью. Внешняя оболочка труба предполагается свободной. Выводятся осесимметричные решения линеаризированных основных уравнений для жидкости и твердого тела, причём жидкость предполагается и вязкой и невязкой. Движение стенки трубы описывается уравнением Доннелла для оболочек. Решение приводит к сложному трансцендентному уравнению распространения волн, содержащему волновое число и частоту. Это уравнение детально исследуется с аналитической и численной точек зрения.

1. Introduction

Current problems in biological modeling, fluidics, and missile fuel systems (propellant lines) have stimulated researchers to study the propagation of waves in elastic or viscoelastic tubes containing liquid at rest or filled with a flowing fluid. Considerable attention is paid, particularly on the part of physiologists concerned with blood flow and the blood pressure pulse in mammalian arteries, to the problem of wave propagation in a system in which the distensibility of the tube is of far greater importance than the compressibility of the liquid. Thus, in a first approximation, the fluid may be regarded as incompressible. There are a number of reasons why it is important to know the precise distribution of the flow velocity, the shear rates, and the mass transport in the blood circulation system. One of the most important is in relation to the onset of atheroma, a degeneration of the arterial wall which leads to the hardening of the arteries. Pathogenic investigations of arteriosclerosis show that the incidence of the disease increases with age, is more pronounced and occurs earlier in male humans, is accelerated by hypertension and prefers certain vessels and locations. Medical evidence indicates that the basic laws of hydraulics determine the behavior of the flow in the arterial tree after loss of elasticity.

The network of vessels containing the blood flow exhibits rather complicated distensibility relations, which in the case of the arteries are of great importance in matching the pumping action of the heart to produce a steady perfusion of the peripheral capillaries. Therefore, the problem that has to be treated is that of pulsatile flow in an elastic vessel.

EULER [1] in 1775 was convinced that ventricular contraction sets up a propagation of waves through the arteries, but he did not succeed in obtaining an expression for the wave velocity. E. H. WEBER [2] announced that he and his brother TH. WEBER had measured the wave velocity in elastic tubes in accordance with the predicted value. Their brother W. WEBER [3] developed a theory of wave transmission, which appeared finally in 1866. Later RESAL [4], KORTEWEG [5] and MOENS [6] worked on the problem also and rediscovered the formula for the velocity of the pulse wave, already found by YOUNG [7] in 1808, namely, that the wave propagation velocity is proportional to the square root of YOUNG's modulus of elasticity of the wall material and the thickness of the wall and inversely proportional to the square root of the liquid density and the diameter of the vessel. According to these investigations, waves of all frequencies are propagated at the same velocity and are not attenuated. This is not true; experimental evidence has shown that the wave velocity increases considerably with increasing frequency of the disturbance.

In connexion with the treatment of the problem, we must remark that the most striking feature of the pulse is its regularity. It is maintained for long periods of time, if no changes occur due to exercise, etc.. Thus it is realistic to regard the arterial pulse as a compound wave with a fundamental frequency, namely that of the heart rate. The arterial circulation is therefore in a steady state of oscillation, and the pressure may be represented by a FOURIER series. Since the average flow velocity of blood is fairly small compared to the propagation speed, a linearization of the equations of fluid motion seems to be justified. In addition, the distension of the arteries is not very large, allowing us to employ linearized shell equations. The diameters of the blood vessels under study are not too small, hence

*) Presented in part at the GAMM Annual Scientific Conference, Wiesbaden, 16–21 April, 1979. An abstract „Über Wellenbewegungen inkompressibler Flüssigkeiten in dünnen elastischen Rohren" appears in ZAMM.

the blood may be considered as a NEWTONian liquid. This is quite a good approximation for the range of higher shear rates which predominate in the larger blood vessels.

Regarding biomechanics in general and flow and pressure in the arterial system in particular, excellent reviews of the state of the art have been given recently [8 through 21]. It is the purpose of the present investigation to make available analytical and numerical results obtained from an extensive study of the dispersion equation [22]. Only the flow in a single tube has been dealt with. This, however, is the basic element out of which, finally, a successful model of the circulatory system will be formulated.

2. Formulation of the problem

Let us consider an infinitely long circular cylindrical tube of radius a and wall thickness h, filled with a streaming homogeneous, viscous and incompressible liquid (Fig. 1). The tube is assumed to be of an elastic and isotropic

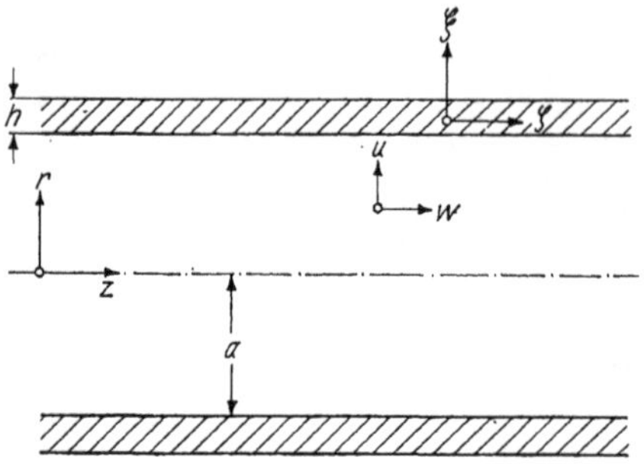

Fig. 1. Fluid flowing through an elastic cylindrical tube, geometry and notations

material. Restricting our attention to a HAGEN-POISEUILLE flow in axial direction, superimposed by small amplitude motions of the liquid and the tube wall, the equations of motion as well as the boundary conditions can be linearized. The small amplitude motion is due to some periodic pressure disturbances. Then the NAVIER-STOKES equations, governing the axisymmetric flow of a viscous NEWTONian fluid, can be put into the following form:

$$\frac{\partial u}{\partial t} + \frac{1}{\varrho}\frac{\partial p}{\partial r} = \nu\left[\frac{\partial^2 u}{\partial r^2} + \frac{1}{r}\frac{\partial u}{\partial r} - \frac{u}{r^2} + \frac{\partial^2 u}{\partial z^2}\right] \tag{1}$$

for the radial direction, and

$$\frac{\partial w}{\partial t} + \frac{1}{\varrho}\frac{\partial p}{\partial z} = \nu\left[\frac{\partial^2 w}{\partial r^2} + \frac{1}{r}\frac{\partial w}{\partial r} + \frac{\partial^2 w}{\partial z^2}\right] \tag{2}$$

for the axial direction. We further have to add the continuity equation

$$\frac{\partial u}{\partial r} + \frac{u}{r} + \frac{\partial w}{\partial z} = 0\,. \tag{3}$$

In these equations, r and z are cylindrical polar coordinates, with r measured in radial direction from the centerline of the tube, and z measured in axial direction along the centerline. The velocity components in direction of increasing r and z, respectively, are denoted by u and w, t is the time.

Moreover, we denote the density of the liquid by ϱ, the kinematic viscosity by ν $(= \mu/\varrho$, with μ as the dynamic viscosity), and the pressure by p.

The motion of the thin-walled tube $(h/a \ll 1)$ can be described by DONNELL's shell equations which, according to HAMMEL [23], may be regarded as the ,,best" first order equations for cylindrical shells. We have

$$\frac{\bar\nu}{a}\frac{\partial \zeta}{\partial z} + \frac{1}{a^2}\xi + \frac{h^2}{12}\frac{\partial^4 \xi}{\partial z^4} + \frac{(1-\bar\nu^2)}{E}\bar\varrho\frac{\partial^2 \xi}{\partial t^2} = \frac{1-\bar\nu^2}{Eh}\left[p - 2\mu\frac{\partial u}{\partial r}\right]_{r=a} \tag{4}$$

in radial direction, and

$$\frac{\partial^2 \zeta}{\partial z^2} + \frac{\bar\nu}{a}\frac{\partial \xi}{\partial z} - \frac{(1-\bar\nu^2)}{E}\bar\varrho\frac{\partial^2 \zeta}{\partial t^2} = \frac{1-\bar\nu^2}{Eh}\mu\left[\frac{\partial u}{\partial z} + \frac{\partial w}{\partial r}\right]_{r=a}, \tag{5}$$

in axial direction. In these equations, axial symmetry has been observed again, ξ and ζ are the displacements of the cylindrical shell in radial and longitudinal direction, respectively. The mass density of the wall material is denoted by $\bar\varrho$, POISSON's ratio by $\bar\nu$ and YOUNG's modulus of elasticity by E. The interaction of the fluid with the solid is expressed by the right hand sides of equations (4) and (5).

At the interface between liquid and wall, $r = a$, the liquid velocity must be equal to the wall velocity, leading to the two kinematic boundary conditions

$$u = \frac{\partial \xi}{\partial t} \quad \text{at} \quad r = a\,, \tag{6}$$

and

$$w = \frac{\partial \zeta}{\partial t} \quad \text{at} \quad r = a \,. \tag{7}$$

The field equations and the boundary conditions (1) through (7) constitute the hydro-elastic problem. Performing the divergence operation on the NAVIER-STOKES equations and taking account of the mass balance equation

$$\operatorname{div} \boldsymbol{v} = 0, \qquad \boldsymbol{v} = \{u, w\}$$

leads to the following expression for the pressure distribution

$$\frac{\partial^2 p}{\partial r^2} + \frac{1}{r}\frac{\partial p}{\partial r} + \frac{\partial^2 p}{\partial z^2} = 0 \,. \tag{8}$$

3. Method of Solution

Since we are interested in wave propagation, and since the coefficients of equations (1) through (8) do not depend on the time (t) and the axial coordinate (z), we may assume for the disturbances a dependency of the form $\exp i(n\omega t \pm k_n z)$ with i as the imaginary unit, n as an integer, ω as the (real) circular frequency, and k_n as the (ecomplex) propagation constant of the n-th harmonic. The pressure is assumed to be given in the form

$$p(r, z, t) = p_0 - \left(\frac{\partial p_0}{\partial z}\right)\bigg|_{z=0} z + \operatorname{Re} \sum_{n=1}^{\infty} \{P^*_{1n}(r)\, e^{-ik_n z} + P^*_{2n}(r)\, e^{ik_n z}\}\, e^{in\omega t} \,, \tag{9}$$

where Re ( ) indicates that the real part of the subsequent expression has to be taken. Substitution of (9) into (8) yields

$$P^*_{jn} = \bar{P}^*_{jn} J_0(ik_n r) \,, \qquad j = 1, 2 \,, \tag{10}$$

with J_0 as the BESSEL function of first kind and zeroth order. Hence, with

$$\bar{P}^*_{jn} = \bar{P}_{jn} - i\bar{\bar{P}}_{jn} \,, \qquad j = 1, 2$$

the pressure can be written as follows

$$p(r, z, t) = p_0 - \left(\frac{\partial p_0}{\partial z}\right)\bigg|_{z=0} z + \sum_{n=1}^{\infty} I_0(k_n r)\, \{[\bar{P}_{1n} \cos(n\omega t - k_n z) + \bar{\bar{P}}_{1n} \sin(n\omega t - k_n z)] +$$
$$+ [\bar{P}_{2n} \cos(n\omega t + k_n z) + \bar{\bar{P}}_{2n} \sin(n\omega t + k_n z)]\} \,. \tag{11}$$

By I_0 we denote the modified BESSEL function of first kind and zeroth order. The index $j = 1$ represents wave propagation along the positive z-axis, while $j = 2$ represents propagation along the negative z-axis. Furthermore, Re k_n is the wave number, Im k_n is the attenuation coefficient, and $c_n = \omega/\text{Re}\, k_n$ is the phase velocity.

For the velocity distribution of the fluid and the displacement of the wall, we seek a solution which is harmonic in time and in the longitudinal space direction, i.e., we put

$$u(r, z, t) = \operatorname{Re} \sum_{n=1}^{\infty} \{U_{1n}(r)\, e^{-ik_n z} + U_{2n}(r)\, e^{ik_n z}\}\, e^{in\omega t} \,, \tag{12}$$

$$w(r, z, t) = W_0(r) + \operatorname{Re} \sum_{n=1}^{\infty} \{W_{1n}(r)\, e^{-ik_n z} + W_{2n}(r)\, e^{ik_n z}\}\, e^{in\omega t} \,, \tag{13}$$

and

$$\xi(z, t) = \xi_0(z) + \operatorname{Re} \sum_{n=1}^{\infty} \{\xi_{1n}\, e^{-ik_n z} + \xi_{2n}\, e^{ik_n z}\}\, e^{in\omega t} \,, \tag{14}$$

$$\zeta(z, t) = \zeta_0(z) + \operatorname{Re} \sum_{n=1}^{\infty} \{\zeta_{1n}\, e^{-ik_n z} + \zeta_{2n}\, e^{ik_n z}\}\, e^{in\omega t} \,. \tag{15}$$

Substitution of (12) and (13) into the continuity equation (3) leads to

$$U'_{jn} + \frac{1}{r} U_{jn} = \pm\, ik_n W_{jn} \,, \tag{16}$$

and

$$\frac{\partial W_0}{\partial z} = 0 \,, \tag{17}$$

where a prime denotes differentiation with respect to r. From the latter equation we conclude that W_0 is a function of r only, a fact already indicated in equation (13). From the NAVIER-STOKES equations (1) and (2), we obtain

$$U''_{jn} + \frac{1}{r} U'_{jn} + \left(\varkappa_n^2 - \frac{1}{r^2}\right) U_{jn} = -\frac{ik_n}{\mu} \bar{P}^*_{jn} J_1(ik_n r) \,, \qquad j = 1, 2 \,, \tag{18}$$

$$W''_{jn} + \frac{1}{r} W'_{jn} + \varkappa_n^2 W_{jn} = \mp \frac{ik_n}{\mu} \bar{P}^*_{jn} J_0(ik_n r) \,, \qquad \begin{cases} j = 1 \,, \\ j = 2 \end{cases} \tag{19}$$

and

$$W_0'' + \frac{1}{r} W_0' = -\frac{1}{\mu}\left(\frac{\partial p_0}{\partial z}\right)_{z=0}, \tag{20}$$

with

$$\varkappa_n^2 = -k_n^2 - \frac{in\omega}{\nu} \tag{21}$$

and J_1 as the BESSEL function of first kind and first order. The solution of equation (20) with $W_0(a) = 0$ reads

$$W_0(r) = \frac{a^2}{4\mu}\left(\frac{\partial p_0}{\partial z}\right)_{z=0}\left[1 - \frac{r^2}{a^2}\right]. \tag{22}$$

This is, however the velocity distribution of the well-known HAGEN-POISEUILLE flow. Equation (19) yields

$$W_{jn}(r) = A_{jn} J_0(\varkappa_n r) \pm \frac{k_n}{\varrho n\omega} \bar{P}_{jn}^* J_0(ik_n r), \qquad \begin{cases} j = 1 \\ j = 2 \end{cases} \tag{23}$$

with A_{jn} as an integration constant. Differentiation of equation (16) with respect to r and substitution of the expression so obtained into equation (18) allows the determination of U_{jn}, namely

$$U_{jn}(r) = \pm \frac{ik_n}{\varkappa_n} A_{jn} J_1(\varkappa_n r) + \frac{ik_n^2}{\varkappa_n^2}\left[\frac{ik_n}{n\varrho\omega} - \frac{1}{\mu k_n}\right] \bar{P}_{jn}^* J_1(ik_n r), \qquad \begin{cases} j = 1 \\ j = 2. \end{cases} \tag{24}$$

Next we turn to the solution of equations (4) and (5), governing the behavior of the shell. Defining the flexural rigidity of the shell by

$$D = \frac{Eh^3}{12(1 - \bar{\nu}^2)} \tag{25}$$

there results, employing equations (14) and (15) as well as (22), (23) und (24), after a few standard manipulations

$$\frac{d^4\xi_0}{dz^4} + \frac{Eh}{Da^2}\xi_0 = \frac{1}{D}\left[p_0 - \frac{2 - \bar{\nu}}{2}\left(\frac{\partial p_0}{\partial z}\right)_{z=0} z\right] + \text{const}, \tag{26}$$

and

$$\xi_{jn}\left\{\frac{1}{a^2} + \frac{h^2}{12}k_n^4 - \frac{1 - \bar{\nu}^2}{E}\varrho n^2\omega^2\right\} \mp \frac{ik_n\bar{\nu}}{a}\zeta_{jn} \pm \frac{2(1 - \bar{\nu}^2)}{Eh} ik_n\mu J_1(\varkappa_n a) A_{jn} +$$

$$+ \frac{(1 - \bar{\nu}^2)}{Eh} \bar{P}_{jn}^*\left\{- J_0(ik_n a) - 2\frac{k_n^2}{\varkappa_n^2}\left[\frac{i\mu k_n^2}{\varrho n\omega} - 1\right] J_1'(ik_n a)\right\} = 0, \tag{27}$$

$$\mp \xi_{jn}\frac{i\bar{\nu}k_n}{a} + \left[\frac{n^2\omega^2}{E}(1 - \bar{\nu}^2)\varrho - k_n^2\right]\zeta_{jn} + \frac{(1 - \bar{\nu}^2)}{Eh}\mu\left(\frac{\varkappa_n^2 - k_n^2}{\varkappa_n}\right) J_1(\varkappa_n a) A_{jn} \pm$$

$$\pm \frac{(1 - \bar{\nu}^2)}{Eh} ik_n J_1(ik_n a)\left[\frac{ik_n}{\varkappa_n^2}\left(\frac{i\mu k_n^2}{\varrho n\omega} - 1\right) + \frac{\mu k_n}{\varrho n\omega}\right]\bar{P}_{jn}^* = 0. \tag{28}$$

In these equations the upper sign belongs to $j = 1$ (waves travelling in the positive z-direction), while the lower sign belongs to $j = 2$ (waves travelling in the negative z-direction). Equation (26) is recognized as the governing differential equation occurring in the general (bending) theory of cylindrical shells loaded symmetrically with respect to its axis. Since solutions are discussed at length by TIMOSHENKO and WOINOWSKY-KRIEGER [24], further details will be discarded. They are irrelevant for the problem under consideration. The boundary conditions (6) and (7) furnish the relations

$$i\omega n\xi_{jn} \mp \frac{ik_n}{\varkappa_n} J_1(\varkappa_n a) A_{jn} - \frac{ik_n}{\mu\varkappa_n^2}\left[\frac{i\mu k_n}{n\omega\varrho} - 1\right] J_1(ik_n a) \bar{P}_{jn}^* = 0, \tag{29}$$

$$i\omega n\zeta_{jn} - J_0(\varkappa_n a) A_{jn} \mp \frac{k_n}{\varrho n\omega} J_0(ik_n a) \bar{P}_{jn}^* = 0. \tag{30}$$

Equations (27) through (30) constitute four linear homogeneous algebraic equations for the unknowns ξ_{jn}, ζ_{jn}, A_{jn} and $\bar{P}_{jn}^*$. A nontrivial solution of this system exists only, if the determinant of the coefficients vanishes.

This postulate, the so-called dispersion relation, leads to a transcendental equation, which may have an infinite number of solutions k_n. In the problem under study, however, only the real values of k_n are of interest, since they represent propagating waves. For pure imaginary or complex k_n-values, the waves are attenuated and decay, depending on the magnitude of $\mathrm{Im}(k_n)$. These waves are non-propagating waves and are of less interest in the present investigation. From the above equations (28) through (30), the ratios $\xi_{jn}/\overline{P}^*_{jn}$, $\zeta_{jn}/\overline{P}^*_{jn}$ and $A_{jn}/\overline{P}^*_{jn}$ can be obtained and introduced into equations (12), (13), (14) and (15). Hence the flow velocities and the wall motion are determined for a given pressure (cf. equation (11)) at $z = 0$. In many practical cases the pressure can only be measured as the mean pressure over the cross-section. Therefore

$$p_{\mathrm{mean}}(z, t) = \frac{2}{a^2} \int_0^a r p(r, z, t)\,\mathrm{d}r = p_0 - \left(\frac{\partial p_0}{\partial z}\right)_{z=0} z +$$

$$+ \frac{2}{a} \sum_{n=1}^{\infty} \frac{1}{k_n} I_1(k_n a) \{[\overline{P}_{1n} \cos(n\omega t - k_n z) + \bar{\bar{P}}_{1n} \sin(n\omega t - k_n z)] +$$

$$+ [\overline{P}_{2n} \cos(n\omega t + k_n z) + \bar{\bar{P}}_{2n} \sin(n\omega t + k_n z)]\} . \tag{31}$$

This expression yields at the location $z = 0$ the value

$$p_{\mathrm{mean}}(0, t) = p_0 + \frac{2}{a} \sum_{n=1}^{\infty} \frac{1}{k_n} I_1(k_n a) \{(\overline{P}_{1n} + \overline{P}_{2n}) \cos n\omega t + (\bar{\bar{P}}_{1n} + \bar{\bar{P}}_{2n}) \sin n\omega t\} , \tag{32}$$

while

$$\frac{\partial p_{\mathrm{mean}}(0, t)}{\partial z} = -\left(\frac{\partial p_0}{\partial z}\right)\bigg|_{z=0} + \frac{2}{a} \sum_{n=1}^{\infty} I_1(k_n a) \{(\bar{\bar{P}}_{2n} - \bar{\bar{P}}_{1n}) \cos n\omega t - (\overline{P}_{2n} - \overline{P}_{1n}) \sin n\omega t\} . \tag{33}$$

I_1 denotes the modified BESSEL function of the first kind and first order.

4. Discussion of the Dispersion Equation

The dispersion relation can be put into the form

$$\Delta = \mathrm{Det}\,(a_{\alpha\beta}) = 0 , \qquad \alpha, \beta = 1, 2, 3, 4 \tag{34}$$

where elements of the determinant are given by

$$a_{11} = \frac{1}{a^2} + \frac{h^2}{12} k_n^4 - \frac{1 - \bar{\nu}^2}{E} \bar{\varrho} n^2 \omega^2 , \qquad a_{12} = \mp \frac{i k_n \bar{\nu}}{a} , \qquad a_{13} = \pm \frac{2(1 - \bar{\nu}^2)}{Eh} \mu i k_n J_1(\varkappa_n a) ,$$

$$a_{14} = -\frac{1 - \bar{\nu}^2}{Eh} \left[J_0(i k_n a) + \frac{2 k_n^2}{\varkappa_n^2} \left(\frac{i k_n^2 \mu}{\varrho n \omega} - 1\right) J_1'(i k_n a)\right] ,$$

$$a_{21} = \mp \frac{i k_n \bar{\nu}}{a} , \qquad a_{22} = \frac{1 - \bar{\nu}^2}{E} \bar{\varrho}\, n^2 \omega^2 - k_n^2 , \qquad a_{23} = \frac{1 - \bar{\nu}^2}{Eh} \mu \frac{\varkappa_n^2 - k_n^2}{\varkappa_n} J_1(\varkappa_n a) ,$$

$$a_{24} = \pm \frac{1 - \bar{\nu}^2}{Eh} i k_n J_1(i k_n a) \left[\frac{i k_n}{\varkappa_n^2} \left(\frac{i \mu k_n^2}{\varrho n \omega} - 1\right) + \frac{\mu k_n}{\varrho n \omega}\right] ,$$

$$a_{31} = i n \omega , \qquad a_{32} = 0 , \qquad a_{33} = \mp \frac{i k_n}{\varkappa_n} J_1(\varkappa_n a) ,$$

$$a_{34} = -\frac{i k_n}{\mu \varkappa_n^2} \left(\frac{i k_n^2 \mu}{\varrho n \omega} - 1\right) J_1(i k_n a) ,$$

$$a_{41} = 0 , \qquad a_{42} = i n \omega , \qquad a_{43} = -J_0(\varkappa_n a) , \qquad a_{44} = \mp \frac{k_n}{\varrho n \omega} J_0(i k_n a) .$$

With

$$J_1'(\varkappa_n a) = J_0(\varkappa_n a) - \frac{1}{\varkappa_n a} J_1(\varkappa_n a) , \qquad J_1'(i k_n a) = J_0(i k_n a) - \frac{1}{i k_n a} J_1(i k_n a)$$

and

$$F_n(\varkappa_n a) := \frac{J_0(\varkappa_n a)}{J_1(\varkappa_n a)} , \qquad G_n(i k_n a) := \frac{J_0(i k_n a)}{J_1(i k_n a)} \tag{35}$$

there follows after straightforward but tediuns calculations

$$\Delta = \text{Det}\,(a_{\alpha\beta}) = \left[\left(\frac{1}{a^2} + \frac{h^2}{12} k_n^4 - \frac{1-\bar{\nu}^2}{E} \varrho n^2\omega^2\right)\left(\frac{1-\bar{\nu}^2}{E}\varrho n^2\omega^2 - k_n^2\right) + \frac{\bar{\nu}^2 k_n^2}{a^2}\right] \times$$

$$\times \left[i\frac{k_n^2}{\varkappa_n \varrho n\omega} G_n(ik_n a) - \frac{ik_n}{\mu\varkappa_n^2}\left(\frac{i\mu k_n^2}{\varrho n\omega} - 1\right) F_n(\varkappa_n a)\right] + \left(i\frac{n\omega}{a^2} + i\frac{n\omega h^2}{12} k_n^4 - i\frac{1-\bar{\nu}^2}{E}\bar{\varrho} n^3\omega^3\right) \times$$

$$\times \left[- i\frac{(1-\bar{\nu}^2)(\varkappa_n^2 - k_n^2)\, k_n}{Eh\varkappa_n^3}\left(\frac{i\mu k_n^2}{\varrho n\omega} - 1\right) - \frac{(1-\bar{\nu}^2)\, k_n^2}{Eh\varkappa_n}\left(\frac{ik_n}{\varkappa_n^2}\left(\frac{i\mu k_n^2}{\varrho n\omega} - 1\right) + \frac{\mu k_n}{\varrho n\omega}\right)\right] \mp$$

$$\mp \frac{n\omega\bar{\nu}k_n}{a}\left[\pm \frac{2(1-\bar{\nu}^2)\, k_n^2}{Eh\varkappa_n^2}\left(F_n(\varkappa_n a) - \frac{1}{a\varkappa_n}\right)\left(\frac{i\mu k_n^2}{\varrho n\omega} - 1\right) \mp\right.$$

$$\left.\mp i\frac{(1-\bar{\nu}^2)\, k_n}{Eh\varkappa_n}\left\{G_n(ik_n a) + \frac{2k_n^2}{\varkappa_n^2}\left(\frac{i\mu k_n^2}{\varrho n\omega} - 1\right)\left(G_n(ik_n a) - \frac{1}{ik_n a}\right)\right\}\right] \pm$$

$$\pm \frac{n\omega\bar{\nu}k_n}{a}\left[\mp \frac{(1-\bar{\nu}^2)(\varkappa_n^2 - k_n^2)\,\mu k_n}{Eh\varkappa_n\varrho n\omega} G_n(ik_n a) \pm i\frac{(1-\bar{\nu}^2)\, k_n}{Eh} F_n(\varkappa_n a)\left\{\frac{ik_n}{\varkappa_n^2}\left(\frac{i\mu k_n^2}{\varrho n\omega} - 1\right) + \frac{\mu k_n}{\varrho n\omega}\right\}\right] -$$

$$- \left(\frac{(1-\bar{\nu}^2)\,\bar{\varrho} n^3\omega^3}{E} - n\omega k_n^2\right)\left[- i\frac{2(1-\bar{\nu}^2)\,\mu k_n^2}{Eh\varrho n\omega} G_n(ik_n a)\left(F_n(\varkappa_n a) - \frac{1}{\varkappa_n a}\right) -\right.$$

$$\left.- \frac{1-\bar{\nu}^2}{Eh} F_n(\varkappa_n a)\left\{G_n(ik_n a) + \frac{2k_n^2}{\varkappa_n^2}\left(\frac{i\mu k_n^2}{\varrho n\omega} - 1\right)\left(G_n(ik_n a) - \frac{1}{ik_n a}\right)\right\}\right] -$$

$$- n^2\omega^2\left[- \frac{2(1-\bar{\nu}^2)^2\,\mu k_n^2}{E^2h^2}\left(F_n(\varkappa_n a) - \frac{1}{\varkappa_n a}\right)\left\{\left(\frac{i\mu k_n^2}{\varrho n\omega} - 1\right)\frac{ik_n}{\varkappa_n^2} + \frac{\mu k_n}{\varrho n\omega}\right\} +\right.$$

$$\left.+ \frac{(1-\bar{\nu}^2)^2(\varkappa_n^2 - k_n^2)\,\mu}{E^2h^2\varkappa_n}\left\{G_n(ik_n a) + \frac{2k_n^2}{\varkappa_n^2}\left(\frac{i\mu k_n^2}{\varrho n\omega} - 1\right)\left(G_n(ik_n a) - \frac{1}{ik_n a}\right)\right\}\right] = 0\,. \qquad (36)$$

A. Viscous incompressible fluid in an elastic tube

First of all we discuss the solution of the dispersion equation (36) for $|k_n| \ll 1$ and small values of ω. This case is of particular interest with respect to applications in hemoelastodynamics. (long wavelength, low frequency assumption). To each root $k(\omega)$ of the dispersion equation there exists a solution which corresponds to a mode of propagation. With the presence of viscosity we have to expect, of course, that even the propagating modes are attenuated to some extend. The propagation constant is no longer a real number, we assume, however, that $|\text{Im}\, k|$ is rather small. Modes for which $|\text{Im}\, k|$ is large decay rapidly with distance along the tube.

For $\omega > (\nu/n)\,|k_n^2|$ we define $m = in\omega/\nu$ and develop the function $\varkappa_n = \varkappa_n(k_n, \omega)$ by means of the TAYLOR-theorem. Thus

$$\varkappa_n = im^{1/2} + \frac{ik_n^2}{2} m^{-1/2} + o(m^{-3/2})\,,$$

$$\varkappa_n^{-1} = - im^{-1/2} + \frac{ik_n^2}{2} m^{-3/2} + o(m^{-5/2})\,,$$

$$\varkappa_n^{-2} = -m^{-1} + k_n^2 m^{-2} + o(m^{-3})\,,$$

and

$$G_n(ik_n a) = -\frac{2i}{a} k_n^{-1} - \frac{ia}{4} k_n + \frac{ia^3}{96} k_n^3 + o(k_n^5)\,, \qquad (|k_n| \ll 1,\ |k_n a| \ll 1)\,,$$

$$F_n(\varkappa_n a) \sim - i\left(\frac{2}{am^{1/2}} + \frac{am^{1/2}}{4}\right) + i\left(\frac{1}{am^{3/2}} - \frac{a}{8m^{1/2}}\right) k_n^2 + o(k_n^4)\,, \qquad (|m| > |k_n^2|,\ |\varkappa_n a| \ll 1)\,.$$

Hence the solution of the dispersion equation takes the form

$$k_n = \left[\frac{f(\omega)}{g(\omega)} + o(k_n^4)\right]^{1/2}, \qquad (37)$$

where $f(\omega)$ and $g(\omega)$ are given by

$$f(\omega) = 8(1-\bar{\nu}^2)\,\varrho n^2\omega^2(4ia\mu + 8ih\nu\bar{\varrho} - \bar{\varrho} ha^2 n\omega)\,, \qquad (38\text{a})$$

$$g(\omega) = (1-\bar{\nu}^2)\, h\bar{\varrho}\varrho a^4 n^3\omega^3 - 20i(1-\bar{\nu}^2)\, h\mu\bar{\varrho} a^2 n^2\omega^2 + 4(1-\bar{\nu}^2)\,\bar{\varrho} h^2 a^3 n^3\omega^3 - 8E\varrho ha^2 n\omega -$$
$$- 80(1-\bar{\nu}^2)\, a\mu^2 n\omega - 4i(1-\bar{\nu}^2)\,\varrho\mu a^3 n^2\omega^2 - 64iE\mu\bar{\nu} + 32(1-\bar{\nu}^2)\,\bar{\varrho}\mu\nu h\omega +$$
$$+ 80iEh\mu - 4Eh^2 a\bar{\varrho} n\omega\,. \qquad (38\text{b})$$

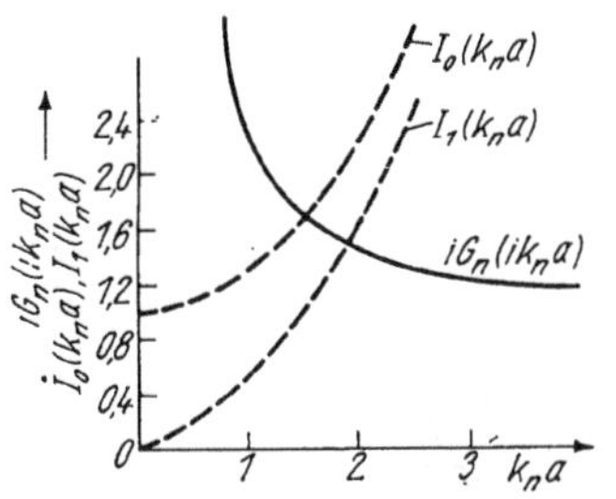

Fig. 2. Plot of $iG_n(ik_na)$, $I_0(k_na)$ and $I_1(k_na)$ vs. k_na

A plot of the real function $iG_n(ik_na)$ vs. k_na (k_n real) is shown in Fig. 2. Note that $iG_n(ik_na) \to \infty$ for $k_na \to 0$, and $iG_n(ik_na) \to 0$ for $k_na \to \infty$. For small values of a and h, equation (37) reduces to

$$k_n \sim \left[\frac{2(1-\bar{\nu}^2)\,\varrho a n^2 \omega^2}{Eh(5-4\bar{\nu})}\right]^{1/2}. \tag{39}$$

With that we obtain for the phase velocity

$$c \sim \left[\frac{5-4\bar{\nu}}{2n^2}\cdot\left(\frac{h}{a}\right)\cdot\left(\frac{\bar{\varrho}}{\varrho}\right)\right]^{1/2} c_L\,, \tag{40}$$

where

$$c_L = \left[\frac{E}{(1-\bar{\nu}^2)\,\bar{\varrho}}\right]^{1/2} \tag{41}$$

denotes the (finite) wavespeed of longitudinal waves in the tube wall.

If $\omega \gg (\nu/n)\cdot|k_{n1}^2|$; there results from the dispersion equation (36)

$$\omega \sim \frac{\nu}{4n}\left[\frac{\bar{\varrho}h+\varrho a}{\bar{\varrho}ah} + o(m^{-1/2})\right]^2, \qquad (|k_n| \ll 1,\ |k_na| \ll 1) \tag{42}$$

and therefore

$$c \sim \frac{\nu}{4nk_{n1}}\left[\frac{\varrho h+\varrho a}{\bar{\varrho}ah}\right]^2, \qquad k_{n1} = \mathrm{Re}\,(k_n)\,. \tag{43}$$

Figures 3–6 represent results of the numerical calculations employing equation (37) and the following typical values of the parameters: $\mu = 0.04$ g cm^{-1} s^{-1}, $\nu = 0{,}038$ cm^2 s^{-1}, $\bar{\varrho} = 1{,}1$ g cm^{-3}, $E = 157 \cdot 10^5$ g cm^{-1} s^{-2}, $\bar{\nu} = 0{,}5$, $a = 0.1$ cm, $h = 0{,}019$ cm and $n = 10$. The graphs labelled by Γ correspond to $h/a = 0{,}19$; $n = 10$; $\bar{\nu} = 0{,}5$ and $\varrho/\bar{\varrho} = 0{,}95$, while the other curves indicate the altered parameter, the others remaining unchanged. The angular frequency ω was made dimensionless by $\omega_0 = [E/\bar{\varrho}a^2(1-\bar{\nu}^2)]^{1/2}$. For very small values of ω, damping of the waves

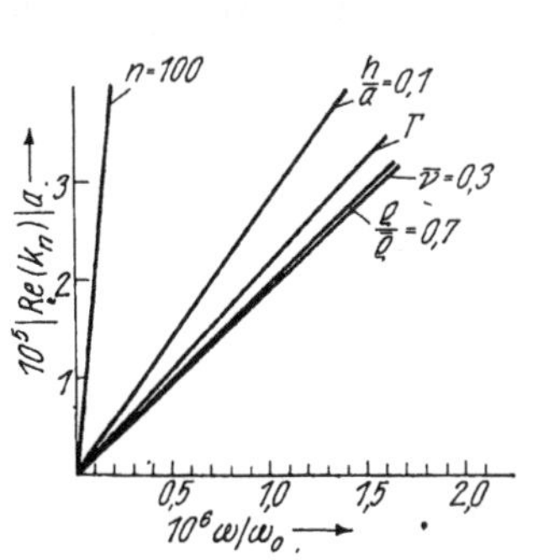

Fig. 3. Plot of $10^5\,|\,\mathrm{Re}(k_n)\,|\,a$ vs. $10^6\,\omega/\omega_0$ based on Eq. (37) for different values of h/a, $\varrho/\bar{\varrho}$, $\bar{\nu}$, and n. The curve Γ corresponds to the standard values

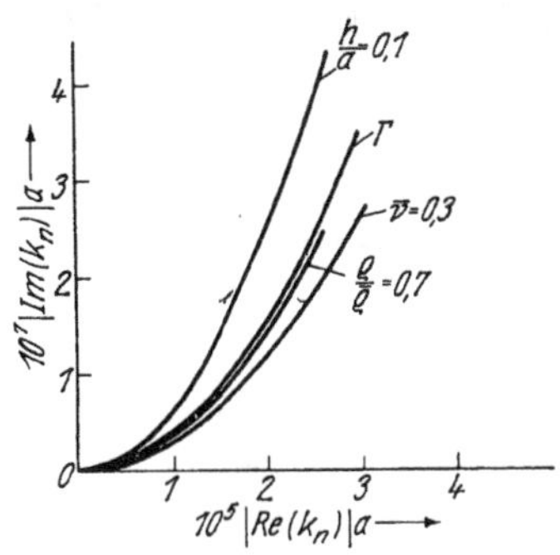

Fig. 4. Plot of $10^7\,|\,\mathrm{Im}(k_n)\,|\,a$ vs. $10^5\,|\,\mathrm{Re}(k_n)\,|\,a$ based on Eq. (37) for different values of h/a, $\varrho/\bar{\varrho}$, $\bar{\nu}$, and n. The curve Γ corresponds to $h/a = 0{,}19$; $\bar{\nu} = 0{,}5$; $\varrho/\bar{\varrho} = 0{,}95$; $n = 10$, 100

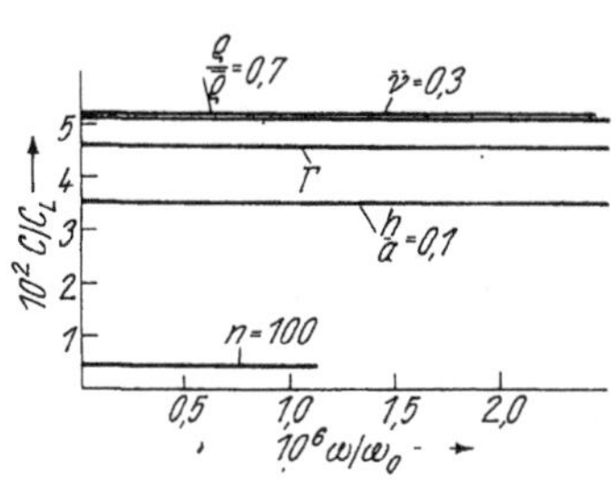

Fig. 5. Plot of $10^2\,c/c_L$ vs. $10^6\,\omega/\omega_0$ based on Eq. (37) for different values of h/a, $\varrho/\bar{\varrho}$, $\bar{\nu}$, and n. The curve Γ corresponds to the standard values

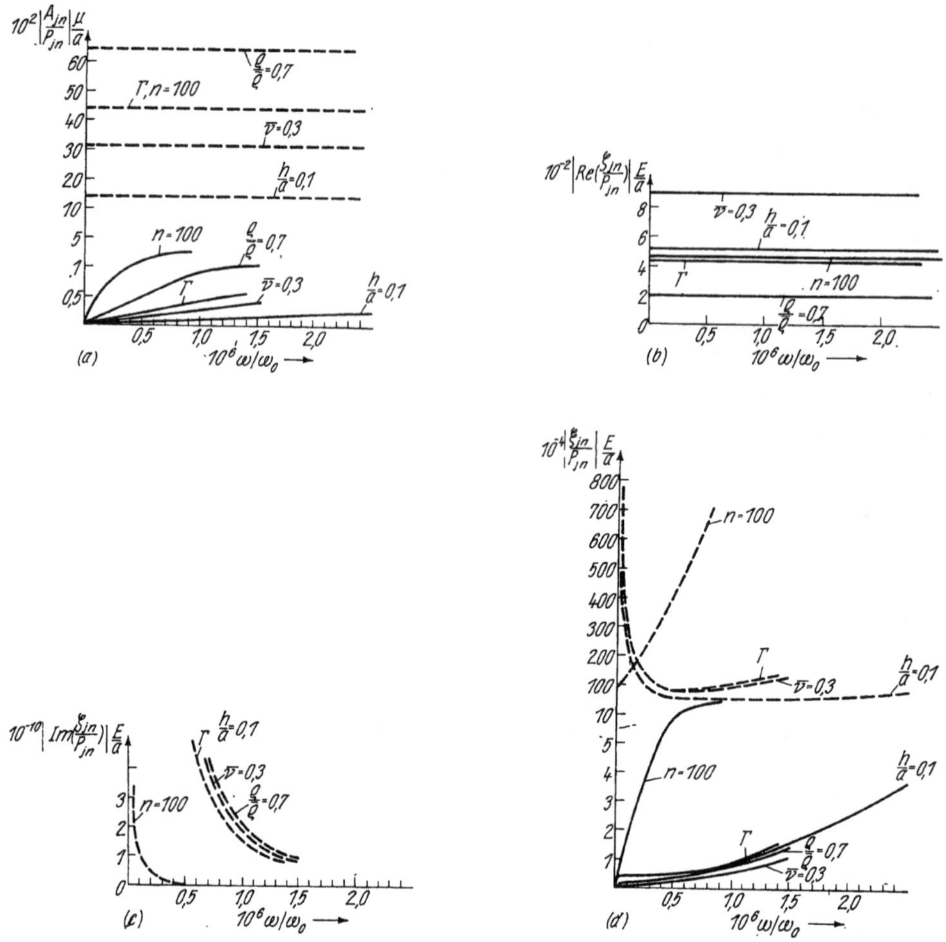

Fig. 6a—6d. Plot of different amplitude ratios vs. $10^6\,\omega/\omega_0$ based on Eqs. (37) and (44) for different values of h/a, $\varrho/\bar{\varrho}$, $\bar{\nu}$, and n. (———): Re (· · · · · ·), (— — —): Im (· · · · · ·). The curve Γ corresponds to the standard values

is relatively small, while for increasing Re (k_n), the attenuation increases rapidly (Fig. 4), i.e. attenuation increase with increasing ω. In case of very small ω, we observe a constant phase velocity (Fig. 5). Figure 6 (a — d), show the different amplitude ratios A_{jn}/P_{jn}, ζ_{jn}/P_{jn} and ξ_{jn}/P_{jn} vs. ω/ω_0, where $P_{jn} \equiv \bar{P}^*_{jn}$. In particular we have

$$\frac{A_{jn}}{P_{jn}} \sim H(k_n, \omega)\,, \tag{44a}$$

$$\frac{\xi_{jn}}{P_{jn}} \sim \mp \frac{k_n}{\nu\left(i\,\frac{n\omega}{\nu}\right)^{3/2}} H(k_n,\omega) - \frac{i k_n}{\varrho n^2 \omega^2}\,, \tag{44b}$$

$$\frac{\zeta_{jn}}{P_{jn}} \sim \frac{2i}{a\nu\left(i\,\frac{n\omega}{\nu}\right)^{3/2}} H(k_n,\omega) \pm \frac{2i}{\varrho a n^2 \omega}\,, \tag{44c}$$

with $j = 1, 2$ and

$$H(k_n, \omega) = \pm \frac{2ih\left[\frac{\bar{\nu}}{ah} + \frac{\bar{\varrho}}{\varrho}\cdot\left(\frac{1}{a^2} - \frac{(1-\bar{\nu}^2)\,\bar{\varrho} n^2\omega^2}{E}\right)\right]}{a\left(\frac{1}{a^2} - \frac{(1-\bar{\nu}^2)\,\varrho n^2\omega^2}{E}\right)\left(\frac{2\bar{\varrho}\nu}{a} - \frac{\mu}{h} + \frac{i\bar{\varrho} a n\omega}{4}\right)} + o(k_n^2)\,. \tag{45}$$

The solid lines correspond to Re $H(k_n, \omega)$, the dashed lines to Im $H(k_n, \omega)$.

Next let us consider the limiting case of a rigid tube ($E \to \infty$). The dispersion equation (36) reads

$$\left[\left(-\frac{1}{a^2} - \frac{h^2}{12}\,k_n^4\right)\cdot k_n^2 + \frac{\bar{\nu}^2 k_n^2}{a^2}\right]\cdot\left[\frac{k_n}{n\omega}\,G_n(ik_n a) - \frac{1}{\nu k_n}\left(\frac{i\nu k_n^2}{n\omega} - 1\right)F_n(\varkappa_n a)\right] = 0\,. \tag{46}$$

This relation is satisfied if one of the brackets vanishes. The first bracket is independent of the kinematic viscosity ν and hence yields solutions for a nonviscous fluid, namely (cf. B)

$$k_n = 0 \quad \text{and} \quad k_n^4 = -\frac{12(1-\bar{\nu}^2)}{a^2h^2}. \tag{47}$$

In both cases no wave propagation occurs. A detailled study of the second bracket reveals that waves are not propagated. This result is confirmed also by numerical analysis [22]. On the other hand it has been demonstrated by RUBINOW and KELLER [17] that in a rigid tube, filled by a compressible nonviscous fluid: waves occur due to compressibility.

Finally, the special cases of waves in a fluid contained in a negligibly thin tube ($h = 0$) and waves in an empty thin tube ($\varrho = 0$) deserve mentioning. In the first case the situation is the same as wave propagation in a fluid jet. There exist waves, however, the propagating modes are strongly attenuated. For waves in the tube alone the dispersion equation (36) can be written as

$$\left[\left(\frac{1}{a^2}+\frac{h^2}{12}k_n^4-\frac{1-\bar{\nu}^2}{E}\bar{\varrho}n^2\omega^2\right)\cdot\left(\frac{1-\bar{\nu}^2}{E}\bar{\varrho}n^2\omega^2-k_n^2\right)+\frac{\bar{\nu}^2k_n^2}{a^2}\right]\times$$
$$\times\left[\frac{k_n}{n\omega}G_n(ik_na)-\frac{1}{\nu k_n}\left(\frac{i\nu k_n^2}{n\omega}-1\right)F_n(\varkappa_na)\right]\frac{ik_n}{\varkappa_n}=0. \tag{48}$$

One solution of this equation is given by $k_n = 0$, since $(k_n/n\omega)\cdot G_n(ik_na)$ remains finite for $k_n \to 0$. With that $c \to \infty$. The expression in the second bracket is identical with the second bracket of the dispersion equation for the limiting case of a rigid tube (cf. eq. (46)). As shown there, waves do not propagate along the tube. The first term of (48) is independent of ν; for $n = 1$, the bracket is identical with the dispersion equation for an inviscid fluid in an "empty"-tube and will be studied to some extend in the next section. It turns out that for real values of k, the values of ω are real too and positive. For $n \to \infty$ we have $\omega = 0$ and therefore $c = 0$. The solutions for $1 < n < \infty$ are

$$\omega^{\pm} = [f_1(k_n) \pm (f_1^2(k_n) - g_1(k_n))^{1/2}]^{1/2}, \tag{49}$$

where

$$f_1(k_n) = \frac{1+\frac{1}{12}h^2a^2k_n^4+a^2k_n^2}{2\,\frac{1-\bar{\nu}^2}{E}\,a^2\bar{\varrho}n}, \qquad g_1(k_n) = \frac{k_n^2(1-\bar{\nu}^2)+\frac{1}{12}h^2a^2k_n^4}{\left(\frac{1-\bar{\nu}^2}{E}\bar{\varrho}n\right)^2}. \tag{50a), (50b}$$

Plots of k_na vs. ω/ω_0, and c/c_L vs. ω/ω_0, based on equation (49), are shown in Figures 7 and 8. They exhibit in particular the role of n, and also of the POISSON number (standard value $\bar{\nu} = 0{,}5$).

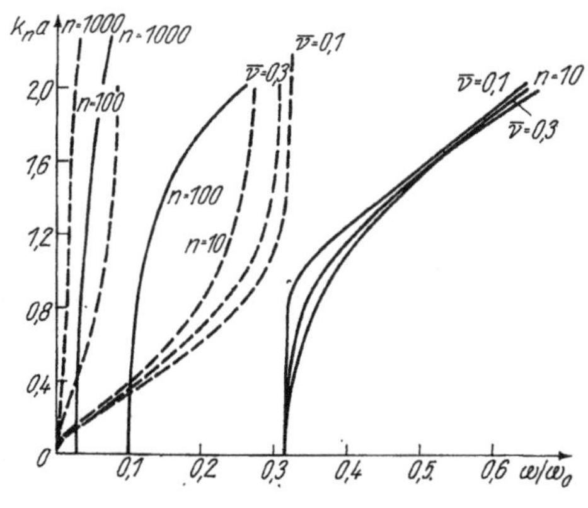

Fig. 7. Plot of k_na vs. ω/ω_0 based on Eq. (49) for different values of $\bar{\nu}$ and n. (———): ω_+^+ solution, (— — —): ω^--solution

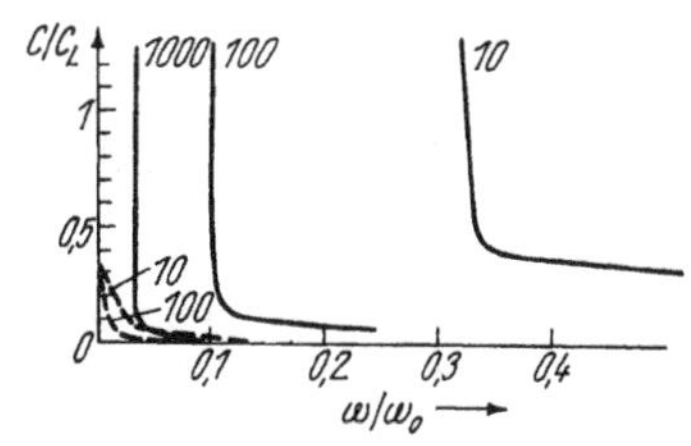

Fig. 8. Plot of c/c_L vs. ω/ω_0 based on Eq. (49) for different values of n

B. Inviscid incompressible fluid in an elastic tube

The next part of the discussion deals with the behavior of an ideal (nonviscous) fluid in an elastic tube. In the absence of waves, the tube is a circular cylindrical shell of constant thickness, and the fluid is at rest. With negligible ν, the NAVIER-STOKES equations are replaced by the EULER equations, while the continuity equation remains unchanged. The DONNELL equations are simplified also, in the first one, the term $2\mu\left(\frac{\partial u}{\partial r}\right)_{r=a}$ is zero, while in the second equation, the right hand side vanishes. Again, let us seek a solution of the linearized equations of motion of the fluid and the shell which is harmonic in time and in the longitudinal space direction:

$$[p, u, w, \xi, \zeta] = [\tilde{p}(r), \tilde{u}(r), \Xi_0, Z_0] \exp(-kz + \omega t). \tag{51}$$

Here Ξ_0 and Z_0 are constants. Substitution in equations (8) and (9) and the Euler equations yields

$$\tilde{p}(r) = P_0 J_0(ikr)\,, \qquad \tilde{u}(r) = \frac{k}{\varrho\omega} P_0 J_1(ikr)\,, \qquad \tilde{w}(r) = \frac{1}{\varrho\omega} P_0 J_0(ikr)\,, \tag{52}$$

where P_0 is an undetermined constant. From the Donnell equations there follows

$$ik\frac{\bar{\nu}}{a} Z_0 + \left(\frac{1}{a^2} + \frac{h^2k^4}{12} - \frac{1-\bar{\nu}^2}{E}\bar{\varrho}\omega^2\right)\Xi_0 = -\frac{1-\bar{\nu}^2}{E} P_0 J_0(ika)\,, \tag{53}$$

$$-k^2 Z_0 - \frac{\bar{\nu}}{a} ik\,\Xi_0 + \frac{1-\bar{\nu}^2}{E}\bar{\varrho}\omega^2 Z_0 = 0\,. \tag{54}$$

Substitution of equation (51) into the boundary condition

$$u = \frac{\partial\xi}{\partial t} \qquad \text{for} \quad r = a\,, \tag{55}$$

and of the resulting relation into equation (52) yields

$$\frac{k}{\varrho\omega} P_0 J_1(ika) = i\omega\Xi_0\,. \tag{56}$$

Now equations (53), (54), and (56) are a linear homogeneous, algebraic system of equations for the unknowns P_0, Ξ_0, and Z_0. In order to obtain a nontrivial solution, the determinant of the matrix of the coefficients must vanish. This leads to the dispersion equation

$$-\omega^4\left(\frac{1-\bar{\nu}^2}{E}\bar{\varrho}\right)^2\left(\frac{\varrho}{\bar{\varrho}}\frac{J_0(ika)}{J_1(ika)}\frac{i}{hk} + 1\right) +$$

$$+\omega^2\left(\frac{1-\bar{\nu}^2}{E}\bar{\varrho}\right)\left(\frac{1}{a^2} + \frac{h^2k^4}{12} + k^2\left(1 + \frac{\varrho}{\bar{\varrho}}\frac{J_0(ika)}{J_1(ika)}\frac{i}{hk}\right)\right) - k^2\left[\frac{1}{a^2} + \frac{h^2k^4}{12} - \left(\frac{\bar{\nu}}{a}\right)^2\right] = 0\,. \tag{57}$$

The solution of this equation can be put into the form

$$\omega^{\pm} = \left[\frac{E}{(1-\bar{\nu}^2\bar{\varrho}a^2)}\right]^{1/2}\{f_0(k) \pm [f_0^2(k) - g_0(k)]^{1/2}\}^{1/2}\,, \tag{58}$$

$$f_0(k) = \frac{1 + \frac{1}{12}h^2a^2k^4 + a^2k^2\left(1 + \frac{\varrho}{\bar{\varrho}}\frac{I_0(ka)}{I_1(ka)}\frac{1}{hk}\right)}{2\left(1 + \frac{\varrho}{\bar{\varrho}}\frac{I_0(ka)}{I_1(ka)}\frac{1}{hk}\right)}\,, \tag{59a}$$

$$g_0(k) = \frac{a^2k^2\left(1 - \bar{\nu}^2 + \frac{1}{12}h^2a^2k^4\right)}{1 + \frac{\varrho}{\bar{\varrho}}\frac{I_0(ka)}{I_1(ka)}\frac{1}{hk}}\,. \tag{59b}$$

From equations (54) and (56), we find for the ratios Ξ_0/Z_0 and P_0/Ξ_0 the relations

$$\frac{\Xi_0}{Z_0} = \frac{a[(1-\bar{\nu}^2)\,\bar{\varrho}\omega^2 - Ek^2]}{i\bar{\nu}Ek}\,, \qquad \frac{P_0}{\Xi_0} = \frac{\varrho\omega^2}{kI_1(ka)}\,. \tag{60}$$

Numerical results according to equations (58) and (60) are plotted in Figures 9—13. Figures 9 and 10 show that $(\omega/\omega_0)^+$ is nearly independent of the parameters (h/a), $(\varrho/\bar{\varrho})$, and $\bar{\nu}$. Furthermore, $(\omega/\omega_0)^-$ increases for decreasing

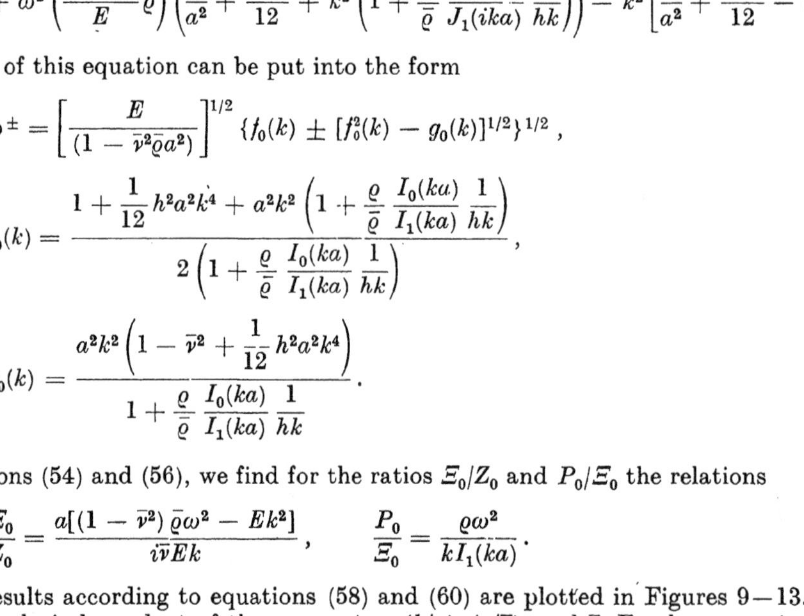

Fig. 9 a, b. Plot of ka vs. ω/ω_0 based on Eq. (58).
(———): ω^+-solution, (— — —): ω^--solution for different values of h/a (Figure a), and $\varrho/\bar{\varrho}$ (Figure b). The (+)-solution is independent of h/a and $\varrho/\bar{\varrho}$. The (−)-solution (a) shows curves with constand h and variable a, the (−)-solution (b) shows curves with constant ϱ and variable $\bar{\varrho}$

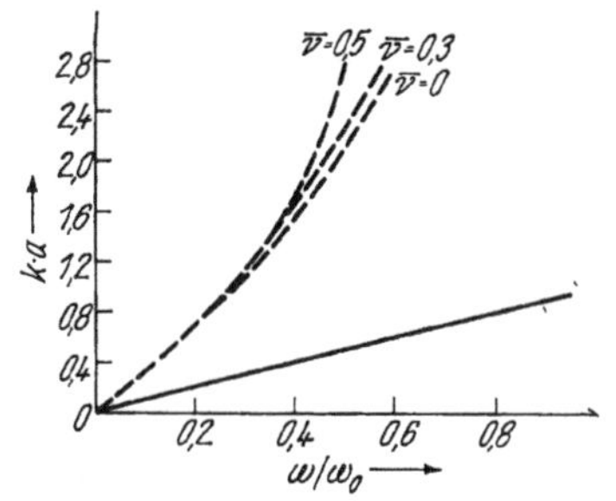

Fig. 10. Plot of ka vs. ω/ω_0 based on Eq. (58) for different values of $\bar{\nu}$.
(———): (+)-solution, (— — —): (−)-solution

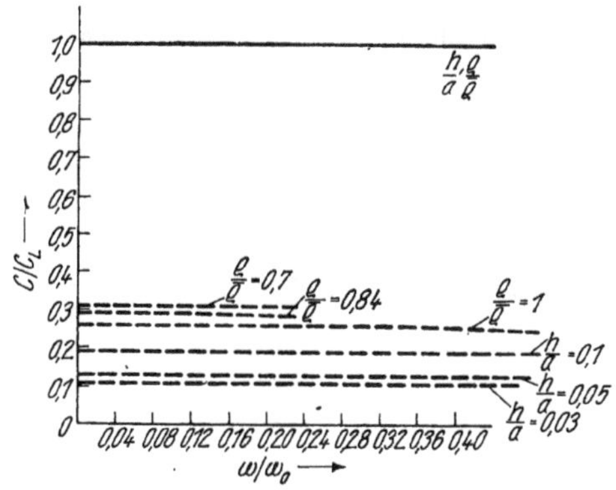

Fig. 11. Plot of c/c_L vs. ω/ω_0 based on Eq. (58) for different values of h/a and $\varrho/\bar{\varrho}$, with constant h and variable a, and constant ϱ and variable $\bar{\varrho}$.
(———): (+)-solution, (— — —): (−)-solution

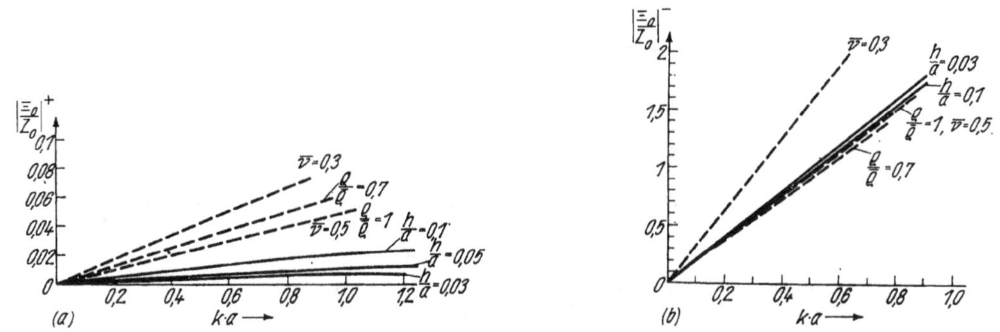

Fig. 12 a, b. Solution of (Ξ_0/Z_0) according to Eq. (60) vs. ka for different values of h/a and $\varrho/\bar{\varrho}$, with constant h and variable a, and constant ϱ and variable $\bar{\varrho}$. The ω^+-solution is given by (a), the (−)-solution by (b). In addition curves are drawn for $\bar{\nu} = 0{,}3$ and $\bar{\nu} = 0{,}5$

Fig. 13 a, b. Solution of $|(P_0\Xi_0)|\,(a/E)$ according to Eq. (60) vs. ka for different values of h/a and $\varrho/\bar{\varrho}$, with constant h and variable a, and constant ϱ and variable $\bar{\varrho}$. The ω^+-solution is given by (a), the ω^--solution by (b)

$(\varrho/\bar{\varrho})$. This has to do with the fact that ω was made dimensionless by ω_0. Again, $(\omega/\omega_0)^-$ increases for decreasing values of $\bar{\nu}$. The dimensionless phase speed $(c/c_L)^+$ is almost constant for small values of ω/ω_0 and nearly independent of (h/a) and $(\varrho/\bar{\varrho})$. Moreover, Figure 11 exhibits that for the standard values of h and a the ratio c/c_L is one, i.e., $c = c_L$. For arterical blood flow we have $|k| \ll 1$, $\omega \ll 1$, and $(\varrho/\bar{\varrho})\,(h/a) \ll 1$. With the different values of $(\varrho/\bar{\varrho})$ and (h/a) we find $0.03 \leqq (\bar{\varrho}h)/(\varrho a) \leqq 0.43$. Comparison with the results reported by RUBINOW and KELLER [17], Figures 18–20*) allow the conclusion that for small ka, or ω/ω_0, respectively, agreement is excellent, while for large values of ka, or ω/ω_0, respectively, the graphs are completely different. However, it should be pointed out that in [17] the authors deal with an inviscid incompressible fluid in a thin elastic tube, the behavior of which is described by the linearized equations of elasticity, while the present study employs DONNELL's equations for a thin elastic shell.

In order to treat analytically the case of small wave numbers (long wavelengths), we observe that with $|ka| \ll 1$, and small values of ω the solution of the dispersion equation is given by

$$\omega^{\pm} = \left[\frac{Ek^2}{\bar{\varrho}(1-\bar{\nu}^2)}\right]^{1/2} \left\{\frac{h\bar{\varrho}+2a\varrho}{4a\varrho} \pm \left[\left(\frac{h\bar{\varrho}+2a\varrho}{4a\varrho^2}\right)^2 - \frac{\bar{\varrho}h(1-\bar{\nu}^2)}{2a\varrho} + o(\omega^4k^2) + o(\omega^2k^4) + o(k^8)\right]^{1/2}\right\}^{1/2}. \quad (61)$$

*) The (−)-curves in [17] correspond to values of ω^+, and the (+)-curves to values of ω^-. Note also that the different values of $\bar{\nu}$ in Fig. 18c are labelled in the wrong order.

For $E \to 0$ (or $\bar{\varrho} \to \infty$, respectively) we find $\omega \to 0$, while for $E \to \infty$, we have $k \to 0$. With $h = 0$, we obtain $\omega = 0$ and $\omega = [E/(1-\bar{\nu}^2)\,\varrho]^{1/2}\,k$, while $\varrho = 0$ (empty tube) leads to

$$k \sim \omega \left(\frac{\bar{\varrho}}{E}\right)^{1/2}, \tag{62}$$

and thus to the constant phase velocity

$$c \sim (1-\bar{\nu})^{1/2}\, c_L\,. \tag{63}$$

Of interest is also the case of a massless tube, yielding

$$\omega \sim \left(\frac{Eh}{2\varrho a}\right)^{1/2} k\,, \tag{64}$$

and therewith

$$c \sim c_Y = \left(\frac{Eh}{2\varrho a}\right)^{1/2}, \tag{65}$$

where c_Y denotes Young's velocity. We see that this velocity depends on both the tube and the fluid properties. It is this velocity which is generally observed in measurements of the pulse velocity in mammalian arteries.

If all the parameters are real it turns out that the inner bracket of equation (61) is real too. Hence for real and positive values of k, the values of ω are real and positive, and ω increases almost linearly with k. In case of $(h/a) \ll 1$ there follows from the solution (61), neglecting the terms $o(\ldots)$,

$$\omega^+ \sim \left[\frac{E}{\bar{\varrho}(1-\bar{\nu}^2)} + o\left(\frac{h}{a}\right)\right]^{1/2} k\,, \qquad \omega^- = o\left(\left(\frac{h}{a}\right)^{1/2} k\right). \tag{66}$$

Since both ω and k are supposed to remain small, we have to postulate that the slope becomes not too large. This condition is satisfied for ω^-, while for ω^+ Young's modulus (E) should be not too large, or the density of the tube material ($\bar{\varrho}$) not too small. Then c^+ approaches the constant wave speed c_L.

Although the dispersion equation (57), together with (68), has only the trivial solution for the problem under consideration, it seems worthwile to study the amplitude ratios. They are for small values of k a good approximation to the exact solutions, which converges toward the zero-solution for $k \to 0$. If $(\varrho/\bar{\varrho}) < (1/2)\,(h/a) \ll 1$, we can discard the second term in the inner bracket of (61), as compared with the first term. Hence

$$\omega^+ \sim \left[\frac{1}{2}\cdot\frac{h}{a}\cdot\frac{E}{\varrho(1-\bar{\nu}^2)} + o\left(\frac{\varrho}{\bar{\varrho}}\right)\right]^{1/2} k\,, \qquad \omega^- \sim o\left(\frac{\varrho}{\bar{\varrho}}\,k\right), \tag{67}$$

and with that

$$c^+ \sim \left[\frac{1}{2}\cdot\frac{h}{a}\cdot\frac{E}{\varrho(1-\bar{\nu}^2)} + o\left(\frac{\varrho}{\bar{\varrho}}\right)\right]^{1/2} \sim c_W, \qquad c^- \sim o\left(\frac{\varrho}{\bar{\varrho}}\right), \tag{68}$$

where

$$c_W = \left[\frac{Eh}{2\varrho(1-\bar{\nu}^2)\,a}\right]^{1/2} \tag{69}$$

is the well known water hammer formula of hydraulics. With equation (67) we find

$$\left(\frac{\Xi_0}{Z_0}\right)^+ \sim i\,\frac{ak}{\nu}\left(1 - \frac{1}{2}\,\frac{h}{a}\,\frac{\bar{\varrho}}{\varrho}\right), \qquad \left(\frac{\Xi_0}{Z_0}\right)^- \sim o(k)\,, \tag{70a}$$

$$\left(\frac{P_0}{\Xi_0}\right)^+ \sim \frac{hE}{a^2(1-\bar{\nu}^2)}\,, \qquad \left(\frac{P_0}{\Xi_0}\right)^- = \text{const}\,. \tag{70b}$$

Finally, a crude estimate of equation (61) allows to set

$$\omega \sim \left[\frac{1}{1-\bar{\nu}^2}\left(\frac{Eh}{2a\varrho} + \frac{E}{\bar{\varrho}}\right)\right]^{1/2} k\,, \tag{71}$$

i.e., in case of small k we recognize that the curves for large and small ω exhibit a similar behavior. The larger $(\varrho/\bar{\varrho})$, as compared with $2ahk^2$ (where $ak \ll 1$), the more accure the approximate formula (71). Furthermore, for very small h the approximation (71) simplifies to (66), except the o-term. With (71), we have

$$c \sim \left[\frac{1}{1-\bar{\nu}^2}\left(\frac{Eh}{2a\varrho} + \frac{E}{\bar{\varrho}}\right)\right]^{1/2}, \tag{72}$$

and, for large values of $\bar{\varrho}$,

$$c \sim c_W\,. \tag{73}$$

This is in agreement with the c^+-formula of (68), but terms of higher order. Together with (71) the ratios Ξ_0/Z_0 and P_0/Ξ_0 are

$$\frac{\Xi_0}{Z_0} \sim i\,\frac{k}{\bar{\nu}}\left(a - \frac{h\bar{\varrho} + 2a\varrho}{2\varrho}\right), \qquad \frac{P_0}{\Xi_0} \sim \frac{E(h\bar{\nu} + 2a)\,h}{2\bar{\varrho}a(1 - \bar{\nu}^2)\,I_1(k_a)}. \tag{74a), (74b}$$

A numerical evalution of the approximate (+)-solution

$$\omega^+ \sim \frac{E}{\bar{\varrho}(1 - \bar{\nu}^2)}\,k\,, \tag{75}$$

valid for $hk \ll 1$, and originating from the dispersion equation (57) for large values of ω, $|k| \gg 1$ and

$$G_n(ik_n a) \sim i - \frac{i}{2a}\,k_n^{-1} - \frac{3i}{8a^2}\,k_n^{-2} - \frac{3i}{8a^3}\,k_n^{-3} - \frac{63}{128a^4}\,k_n^{-4} - \frac{27i}{32a^5}\,k_n^{-5} + o(k_n^{-6})\,, \qquad k_n a \gg 1 \tag{76}$$

shows excellent agreement with the exact numerical solution of the dispersion equation. Practically both curves coincide. The same is true regarding the approximate ω^+-solution (66).

5. Conclusions

A theoretical analysis of the propagation of pressure waves through an infinitely long, thin-walled, elastic tube of circular cross-section is presented. The tube is filled with a streaming incompressible viscous liquid or with an inviscid incompressible liquid, the latter being at rest in the absence of waves. The analysis is restricted to waves whose amplitudes are small. Of special physiological interest (pulsating blood flow through an artery) are waves whose wavelengths are large compared to the radius of the tube. They are discussed in some detail. Analytical expressions are derived and graphs are given which show the dependence of the phase velocity on the frequency for various sets of parameter values. The most important results are summarized as follows:

In case of a rigid tube ($E \to \infty$) there are no waves, regardless whether the liquid is nonviscous or viscous, while in the opposite case ($E = 0$) there exist waves in an inviscid fluid for certain constant values of the wave number (k_n), such that the wave speed increases linearly with the frequency (ω). In a viscous fluid there are waves too, they are, however, strongly attenuated. The same behavior is observed for a viscous liquid jet ($h = 0$), while there are no waves in a nonviscous fluid. Concerning the case of an empty tube ($\varrho = 0$) we find that waves are propagated along the tube wall. With increasing frequency the phase velocity of these waves approaches either (inviscid fluid) the longitudinal wavespeed (c_L), or (viscous fluid) becomes infinite. For small frequencies the complete problem yields a constant phase velocity for both nonviscous or viscous fluids. For increasing values of the frequency damping of the waves in a viscous fluid increases also while in an inviscid fluid waves are propagated without attenuation.

References

1 EULER, L., Principia pro motu sanguinis per arterias determinando, Opera posthuma mathematica et physica anno 1844 detecta, ediderunt P. H. Fuss et N. Fuss, Petropoli, Apud Eggers et socios, Vol. 2, pp. 814–823 (1862).

2 WEBER, E. H., Über die Anwendung der Wellenlehre auf die Lehre vom Kreislaufe des Blutes und insbesondere auf die Pulslehre, Ber. Verh. Kgl. Sächs. Ges. Wiss., Math.-Phys. Kl. 1850.

3 WEBER, W., Theorie der durch Wasser oder andere incompressible Flüssigkeiten in elastischen Röhren fortgepflanzte Wellen, Ber. Verh. Sächs. Ges. Wiss., Math.-Phys. Kl. 18, p. 353–357 (1866).

4 RESAL, H., Sur les petits mouvements d'un fluid incompressible dans un tuyau élastique, C.R. Acad. Sci. 82, 698–699 (1876).

5 KORTEWEG, D. J., Über die Fortpflanzungsgeschwindigkeit des Schalles in elastischen Röhren, Anm. Phys. Chem. (Neue Folge 5, 525–542 (1878).

6 MOENS, A. I., Die Pulskurve, E.J. Brill, NV, Leyden, 1878.

7 YOUNG, TH., Hydraulic investigations, subservient to an intended Croonian lecture on the motion of the blood, Phil. Trans. Roy. Sci. London 88, 164–186 (1808).

8 WOMERSLEY, J. R., An elastic tube theory of pulse transmission and oscillatory flow in mammalian arteries, Wright Air Development Center Tech. Rep. no. 56–684.

9 ATTINGER, E. O. (ed.), Pulsatile Blood Flow, Mc Graw Hill, New York, 1964.

10 MC DONALD, D. A., Blood Flow in Arteries, Second Edition 1974, Williams and Wilkins, Baltimore, Maryland, Edward Arnold (Publishers) Ltd., London.

11 FUNG, Y. C. (ed.), Biomechanics, Am. Soc. Mech. Eng., New York, 1964.

12 ASME — Biomedical Fluid Mechanics Symposium, Am. Soc. Mech. Eng., New York, 1966.

13 FUNG, Y. C., Biomechanics, its scope, history and some problems of continuum mechanics in physiology, Appl. Mech. Rev. 21, 1–20 (1968).

14 COPELEY, A. L. (ed.), Hemorheology, Pergamon Press, Oxford, 1968.

15 FUNG, Y. C. et al. (ed.), Biomechanics, Its Foundations and Objectives, Prentice Hall, Englewood Cliffs., New Jessey, 1971.

16 FUNG, Y. C., Biomechanics: A Survey of the Blood Flow Problem, in Advances in Applied Mechanics, edited by C. S. YIH, Vol. 11, 1971, Academic Press.

17 RUBINOV, S. I.; KELLER, J. B., Wave Propagation in a fluid-filled tube, J. Acoust. Soc. Am. 50, 198–223 (1971).

18 LIGHTHILL, J., Physiological Fluid Mechanics, CISM, Courses and Lectures, No. 111, Udine 1971, Springer, Wien 1971.

19 LIEBERSTEIN, H. M., Mathematical Physiology (Blood Flow and Electrically Active Cells), American Elsevier Publishing Company, Ini., New York, 1973.

20 Lighthill, J., Mathematical Biofluiddynamics, Chapters 10, 12, 13, SIAM 1975.
21 Rubinov, S. I.; Keller, J. B., Wave propagation in a viscoelastic tube containing a viscous fluid, J. Fluid Mech. Vol. 88, part, 1, pp. 181–203 (1978).
22 Metten, St., Über pulsierende Strömungen in elastischen Röhren, Diplomarbeit TH Darmstadt, 1978 (unpublished).
23 Hammel, J., Grundgleichungen der Kreiszylinderschale, Z. Flugwiss. Weltraumf. 1 (1977) Heft 5, S. 353–358.
24 Timoshenko, S. P., Woinowsky-Krieger, S., Theory of Plates and Shells, Second Edition, 1959, Mc Graw Hill, New York.

Eingereicht am 27. 7. 1979

Anschriften: Prof. Dr. rer. nat. H. F. Bauer, Hochschule der Bundeswehr München, Fachbereich Luft- und Raumfahrttechnik, Werner-Heisenberg-Weg 39, D 8014 Neubiberg — Dipl.-Ing. St. Metten, Im Klarenpesch 13, D 502 Frechen — Prof. Dr.-Ing. J. Siekmann, Universität Essen — Gesamthochschule, Fachbereich 12 — Mechanik, Schützenbahn 70, D 4300 Esssen 1 — BRD

ZAMM 60, 235 – 240 (1980)

K. BEYER

Zur Stabilität einer ferromagnetischen Flüssigkeit in einem vertikalen Magnetfeld

Herrn Prof. HERBERT BECKERT zum 60. Geburtstag

Im Rahmen der Variationsrechnung untersuchen wir die Stabilität einer ferromagnetischen Flüssigkeit in einem äußeren Magnetfeld. Für die zweite Variation der Gesamtenergie im ungestörten Ausgangszustand wird eine Ungleichung vom Gårdingschen Typ hergeleitet, die eine gute Einsicht in den Mechanismus des Stabilitätsverlusts bietet. Für unsere Betrachtungen besonders wichtig ist die Tatsache, daß sich die zweite Variation der Gesamtenergie stets in überschaubarer Weise aus dem Systemzustand berechnen läßt.

In this paper we describe a variational approach to the stability theory of a ferromagnetic fluid in an exterior magnetic field. For the second variation of total energy in the unperturbed initial state a Gårding-type inequality is derived, which gives insight in the loss-of-stability mechanism. For our considerations it is essential to have an easy possibility for constructing the second variation of total energy from the underlying state.

В рамках вариационного исчисления изучается устойчивость ферромагнитной жидкости во внешнем магнитном поле. Для второй вариации полной энергии в невозмущённом исходном состоянии выводится неравенство типа Гординга, которое даёт представление о механизме потери устойчивости. Для наших рассмотрений особенно важен тот факт, что вторая вариация полной энергии может быть рассчитана в обозримой форме из состояния системы.

In dieser Arbeit untersuchen wir die Stabilitätsverhältnisse in einer ferromagnetischen Flüssigkeit unter dem Einfluß eines vertikalen äußeren Magnetfeldes. Neben dem ungestörten horizontalen Gleichgewichtszustand der Flüssigkeit existieren bei Überschreiten einer gewissen kritischen Feldstärke noch weitere Gleichgewichtskonfigurationen, die längs der Oberfläche rechteckige bzw. hexagonale Muster bilden. Klassisches Beispiel einer derartigen Erscheinung ist das BÉNARD-Problem (s. [5] und die dort zitierte Literatur). Für eine unendlich tiefe Flüssigkeit mit magnetischer Permeabilität $\mu \sim 1$ wurden die Amplituden der rechteckigen und hexagonalen Wellen in [1] näherungsweise auf der Grundlage eines Variationsprinzips (minimale potentielle Energie) berechnet. Wie schon von LUKE bemerkt (s. [4]), lassen sich auch permanente Schwerewellen längs einer inkompressiblen Flüssigkeit durch ein ähnliches Variationsprinzip beschreiben.

Wir werden uns hier im wesentlichen mit den linearisierten Variationsgleichungen zur potentiellen Energie befassen und für ihre zweite Variation im ungestörten Zustand eine Ungleichung vom GARDINGschen Typ herleiten, die einen guten Einblick in den Mechanismus des Stabilitätsverlusts gestattet. Besonders wichtig für unseren Weg ist es, daß sich die zweite Variation der Gesamtenergie in übersichtlicher Weise stets aus dem Systemzustand berechnen läßt. Dieser Umstand soll an anderer Stelle zur tatsächlichen Lösungskonstruktion benutzt werden.

1.

Wir betrachten die Trennfläche Γ zwischen einer ruhenden inkompressiblen Flüssigkeitsschicht der Dichte ϱ und einem Vakuum unter dem Einfluß eines vertikalen Magnetfelds $\mathcal{H}$ und des Schwerefelds $(0, 0, -g)$. Ist $\mathcal{H} = 0$, so ist die horizontale Trennebene einziger Gleichgewichtszustand. Bei wachsender Feldstärke treten hierzu weitere Gleichgewichtskonfigurationen Γ: $z = \zeta(x, y)$; ihre Form ist aus der längs Γ zu fordernden Bedingung

$$-\beta\,\Delta\zeta + \varrho g\zeta + \frac{1-\mu}{4\pi}\left(\frac{1}{2}\,|\mathcal{H}_t^-|^2 + \mu|\mathcal{H}_n^-|^2\right) = \text{const}\,. \tag{1.1}$$

zu bestimmen. $\beta > 0$ ist hierbei der Koeffizient der Oberflächenspannung der Flüssigkeit und $\mu > 1$ ihre magnetische Permeabilität, $\mathcal{H}_t^-$ bzw. $\mathcal{H}_n^-$ bezeichnen die tangentielle bzw. normale Komponente des Magnetfelds in der Flüssigkeit. Dieser Schreibweise entsprechend werden wir die sich auf das Vakuum ($\mu = 1$) beziehenden Feldgrößen stets mit dem oberen Index „+" kennzeichnen. (1.1) ergibt sich aus den längs Γ gültigen Gleichgewichtsbedingungen

$$(\sigma^+ - \sigma^-)\,n + \beta\,\Delta\zeta n = 0$$ [1])

$-n$ ist der Normalenvektor an Γ und $\sigma = (\sigma_{ik})$

$$\sigma_{ik}^+ = \frac{1}{4\pi}\left(\mathcal{H}_i^+\mathcal{H}_k^+ - \frac{|\mathcal{H}^+|^2}{2}\,\delta_{ik}\right),$$

$$\sigma_{ik}^- = \frac{\mu}{4\pi}\left(\mathcal{H}_i^-\mathcal{H}_k^- - \frac{|\mathcal{H}^-|^2}{2}\,\delta_{ik}\right) - \varrho g z \delta_{ik}$$

der Spannungstensor des Magnetfelds — bei Beachtung der Stetigkeit der tangentiellen Komponente von $\mathcal{H}$ und der normalen der magnetischen Induktion. Im übrigen wird sich (1.1) aus dem zu stellenden Variationsproblem in natürlicher Weise ergeben.

[1]) Vgl. [2], S. 167.

In dieser Arbeit beschränken wir uns auf die Betrachtung periodischer Lösungen der obigen Aufgabe. Wir werden demgemäß von den Lösungen 2π-Periodizität in x- und y-Richtung fordern und die folgenden Überlegungen auf den Quader

$$\Omega = Q \times \{z \mid -z_0 < z < z_1\}, \qquad Q: 0 < x, y < 2\pi \qquad (z_0, z_1 > 0)$$

beziehen. Im feldfreien Zustand nehme die Flüssigkeit hiervon den Raum $-z_0 \leqq z \leqq 0$ ein. Die zu bestimmenden nichttrivialen Gleichgewichtskonfigurationen werden durch 2π-periodische Trennflächen Γ: $z = \zeta(x, y)$ charakterisiert. Diese zerlegen Ω in Teilgebiete Ω^-: $-z_0 < z < \zeta(x, y)$ und Ω^+: $\zeta(x, y) < z < z_1$ (Vakuum). Für willkürlich vorgegebenes Γ notieren wir die Gesamtenergie des Systems:

$$E(\zeta) = \frac{\beta}{2} \int_Q |\nabla\zeta|^2 \, dx \, dy + \frac{\varrho g}{2} \int_Q \zeta^2 \, dx \, dy + \frac{1}{8\pi} \int_{\Omega^+} |\mathcal{H}^+|^2 \, dV + \frac{\mu}{8\pi} \int_{\Omega^-} |\mathcal{H}^-|^2 \, dV, \tag{1.2}$$

$dV = dx \, dy \, dz$. In (1.2) ist das Magnetfeld in Abhängigkeit von ζ an noch zu beschreibende Randbedingungen zu binden. Gleichgewichtszustände sind aus dem Verschwinden der ersten Variation der Gesamtenergie zu bestimmen: $\langle E'(\zeta), \delta\zeta\rangle = 0$ für alle zulässigen Variationen $\delta\zeta$. Naturgemäß hat man sich auf hinreichend reguläre Lösungen bzw. Variationen zu beschränken, etwa $\zeta, \delta\zeta \in C^{2+\alpha}(Q)$. Ich werde aber hier auf diese wichtige Seite nicht eingehen und stattdessen bei den folgenden Überlegungen stets hinreichende Regularität der beteiligten Funktionen annehmen.

Der magnetische Anteil der Gesamtenergie ist durch Randbedingungen für das in (1.2) eingehende Magnetfeld festzulegen. Wir setzen $\mathcal{H}^\pm = H\nabla\psi^\pm$ und suchen die Potentiale $\psi^\pm$ in der Nähe eines vertikalen Parallelfeldes

$$\psi^+ = z + u^+, \qquad \psi^- = \frac{z}{\mu} + u^-, \tag{1.3}$$

$$\Delta u^+ = 0 \quad \text{in } \Omega^+, \qquad \Delta u^- = 0 \quad \text{in } \Omega^-, \tag{1.4}$$

wobei wir

$$u^+(x, y, z_1) = u^-(x, y, -z_0) = 0 \quad \text{für } (x, y) \in Q \tag{1.5}$$

verlangen. Längs Γ haben die Sprungbedingungen $\mathcal{H}_t^+ = \mathcal{H}_t^-$, $\mathcal{H}_n^+ = \mu\mathcal{H}_n^-$ zu gelten; sie führen auf

$$u^+ - u^- = \frac{1-\mu}{\mu} z + \text{const.}, \tag{1.6}$$

$$\frac{\partial u^+}{\partial n} - \mu \frac{\partial u^-}{\partial n} = 0 \quad \text{längs } \Gamma. \tag{1.7}$$

Wir vereinbaren, mit n stets den nach Ω^+ gerichteten Normaleneinheitsvektor an Γ zu bezeichnen. Neben den Periodizitätsforderungen

$$u^\pm(0, y, z) = u^\pm(2\pi, y, z), \qquad u^\pm(x, 0, z) = u^\pm(x, 2\pi, z), \tag{1.8}$$

$$u_x^\pm(0, y, z) = u_x^\pm(2\pi, y, z), \qquad u_y^\pm(x, 0, z) = u_y^\pm(x, 2\pi, z) \tag{1.9}$$

normieren wir das Feld durch die Vorgabe seines Gesamtflusses, bzw.

$$\int_\Gamma \frac{\partial u^+}{\partial n} \, d\sigma = 0; \tag{1.10}$$

$d\sigma$ ist hierbei das Oberflächenelement von Γ.

Man bemerkt, daß die Lösungen des linearen Randwertproblems (1.4)—(1.10) Extremalen des Variationsproblems

$$\int_{\Omega^+} |\nabla u^+|^2 \, dV + \mu \int_{\Omega^-} |\nabla u^-|^2 \, dV \to \text{Min}. \tag{1.11}$$

unter der Nebenbedingung (1.6) sind; die Konstante in (1.6) ist dabei frei variierbar. Bezeichnen wir mit V^+ bzw. mit V^- die SOBOLEWräume der periodischen[1]) Funktionen u^+ bzw. u^- mit über Ω^+ bzw. Ω^- quadratisch summierbaren ersten Ableitungen und Randwerten Null längs der Deck- bzw. Grundfläche von Ω, so ist (1.11) über den Produktraum $V^+ \times V^-$ zu stellen. Man entnimmt dies den für Variationen $(\varphi^+, \varphi^-) \in V^+ \times V^-$, $\varphi^+ - \varphi^- = \text{const.}$ längs Γ, gültigen Variationsgleichungen zu (1.11):

$$\int_{\Omega^+} \nabla u^+ \nabla\varphi^+ \, dV + \mu \int_{\Omega^-} \nabla u^- \nabla\varphi^- \, dV = 0 \tag{1.12}$$

in Verbindung mit dem nachfolgenden einfachen Hilfssatz 1. Danach existiert bei Gültigkeit von (1.12) ein Randfunktional $f \in H^{-1/2}(\Gamma)$ (zu diesen Räumen s. [3]) mit den Eigenschaften

$$\int_{\Omega^+} \nabla u^+ \nabla\varphi^+ \, dV + \mu \int_{\Omega^-} \nabla u^- \nabla\varphi^- \, dV + \int_\Gamma f(\varphi^+ - \varphi^-) \, d\sigma = 0 \tag{1.13}$$

---

[1]) Es ist lediglich (1.8) zu fordern.

für alle $(\varphi^+, \varphi^-) \in V^+ \times V^-$ und

$$\int_\Gamma f \, d\sigma = 0 . \tag{1.14}$$

Nach partieller Integration von (1.13) folgt

$$-\frac{\partial u^+}{\partial n} + f = 0 , \quad \mu \frac{\partial u^-}{\partial n} - f = 0 \quad \text{längs } \Gamma , \tag{1.15}$$

d. h. (1.7). Aus (1.15) folgt (1.10) wegen (1.14). Schließlich sind (1.9) die natürlichen Randbedingungen zu (1.11) längs der Seitenflächen von Ω.

Es bleibt Hilfssatz 1 nachzutragen.

Hilfssatz 1. *T sei eine stetige lineare Abbildung eines* BANACH*raumes X auf einen weiteren* BANACH*raum Y und Y_0 ein linearer Teilraum von Y. Ist x^* ein stetiges lineares Funktional über X mit der Eigenschaft*

$$x^* \in T^{-1}(Y_0)^\perp ,$$

so existiert ein über Y stetiges lineares Funktional y^* mit den Eigenschaften

$$x^* = T^* y^* , \qquad y^* \in Y_0^\perp .$$

Beweis. Wegen $N(T)$ (= Nullraum von T) $\subseteqq T^{-1}(Y_0)$ gilt nach Voraussetzung insbesondere

$$x^* \in N(T)^\perp = R(T^*) \ (= \text{Wertevorrat von } T^*) .$$

Dies zeigt die Existenz von y^* mit $x^* = T^* y^*$. Ist nun $y_0 \in Y_0$ — etwa $y_0 = Tx_0$, $x_0 \in T^{-1}(Y_0)$ —, so muß aber

$$\langle y^*, y_0 \rangle = \langle y^*, Tx_0 \rangle = \langle T^* y^*, x_0 \rangle = \langle x^*, x_0 \rangle = 0$$

zutreffen, also $y^* \in Y_0^\perp$.

Wir haben nunmehr die nach (1.11) eindeutig bestimmte Lösung (u^+, u^-) des Randwertproblems (1.4)—(1.10) in den Magnetfeldanteil E_m der Gesamtenergie einzutragen. Man erhält wegen (1.3) zunächst

$$\frac{4\pi}{H^2} E_m = \frac{1}{2} \int_{\Omega^+} |\nabla u^+|^2 \, dV + \int_{\Omega^+} \nabla z \, \nabla u^+ \, dV + \frac{\mu}{2} \int_{\Omega^-} |\nabla u^-|^2 \, dV + \int_{\Omega^-} \nabla z \, \nabla u^- \, dV + \text{const.} . \tag{1.16}$$

Um der Inkompressibilität der Flüssigkeit zu genügen, werden wir die Lösungsflächen ζ der zusätzlichen Bedingung

$$\int_Q \zeta \, dx \, dy = 0 \tag{1.17}$$

zu unterwerfen haben. In diesem Fall wird die in (1.16) eingehende Konstante unabhängig von ζ und bleibt daher im Folgenden unberücksichtigt. Integrieren wir die linearen Terme in (1.16) partiell, so erhalten wir bei Beachtung sämtlicher Randbedingungen schließlich

$$\frac{4\pi}{H^2} E_m = -\frac{1}{2} \int_{\Omega^+} |\nabla u^+|^2 \, dV - \frac{\mu}{2} \int_{\Omega^-} |\nabla u^-|^2 \, dV . \tag{1.18}$$

Unter dem Einfluß des fixierten Vertikalfelds nimmt E_m also einen maximalen Wert an.

2.

In diesem Abschnitt berechnen wir die Variation von E_m in Abhängigkeit von ζ. Betrachten wir neben der obigen Konfiguration eine Schar von Nachbarflächen Γ_t: $z = \zeta(x, y) + th(x, y)$. Γ_t zerlege Ω in die Teilgebiete Ω_t^+ und Ω_t^-; es ist $\Omega_0^\pm = \Omega^\pm$. Das wie oben zu konstruierende Magnetfeld zu Γ_t sei $u^\pm(t; x, y, z)$. Die partiellen Ableitungen $\dot{u}^\pm$ der Funktionen $u^\pm$ nach der Variablen t werden wir für $t = 0$ abkürzend mit $v^\pm$ benennen: $v^\pm(x, y, z) = \dot{u}^\pm(0; x, y, z)$. Wie die Felder $u^\pm$ hängen die Funktionen $v^\pm$ von der Konfiguration ζ ab, dazu tritt die zusätzliche Abhängigkeit von ihrer Variation h. Wir kennzeichnen diese Abhängigkeiten nicht ausdrücklich, zumal sie sich aus dem jeweiligen Zusammenhang ergeben.

Wir erhalten die erste Variation von E_m, indem wir die Volumenintegrale in (1.18) an der Stelle $t = 0$ differenzieren:

$$\frac{4\pi}{H^2} \langle E_m'(\zeta), h \rangle = \frac{4\pi}{H^2} \frac{d}{dt} E_m(\zeta + th)_{t=0} =$$

$$= -\int_{\Omega^+} \nabla u^+ \, \nabla v^+ \, dV + \frac{1}{2} \int_\Gamma |\nabla u^+|^2 h \, dx \, dy - \mu \int_{\Omega^-} \nabla u^- \, \nabla v^- \, dV - \frac{\mu}{2} \int_\Gamma |\nabla u^-|^2 h \, dx \, dy .$$

Dies läßt sich bei Beachtung von (1.13)—(1.15) sowie der aus (1.6) durch Differentiation hervorgehenden Beziehung

$$v^+ - v^- = -(u_z^+ - u_z^-) h + \frac{1-\mu}{\mu} h + \text{const.} \quad \text{längs } \Gamma \tag{2.1}$$

wie folgt umformen:

$$\frac{4\pi}{H^2}\langle E'_m(\zeta), h\rangle = \frac{1}{2}\int\limits_\Gamma (|\nabla u^+|^2 - \mu\,|\nabla u^-|^2)\,h\,\mathrm{d}x\,\mathrm{d}y + \int\limits_\Gamma f(v^+ - v^-)\,\mathrm{d}\sigma =$$

$$= \frac{1}{2}\int\limits_\Gamma (|\nabla u^+|^2 - \mu\,|\nabla u^-|^2)\,h\,\mathrm{d}x\,\mathrm{d}y + \int\limits_\Gamma \frac{\partial u^+}{\partial n}\left(-u_z^+ + u_z^- + \frac{1-\mu}{\mu}\right) h\,\mathrm{d}\sigma\,. \tag{2.2.}$$

Berücksichtigt man jetzt den Ansatz (1.3), so findet man nach einer einfachen Rechnung in Einklang mit (1.1)

$$\langle E'_m(\zeta), h\rangle = \frac{1-\mu}{4\pi_\mu}\int\limits_\Gamma \left(\frac{1}{2}\,|\mathcal{H}_t^-|^2 + \mu\,|\mathcal{H}_n^-|^2\right) h\,\mathrm{d}x\,\mathrm{d}y\,.$$

Aus (2.2) läßt sich nun die zweite Variation von E_m bestimmen. Wir notieren ihren Wert hier nur für $\zeta = 0$. In diesem Fall ist $u^+ = 0$, $u^- = 0$, demzufolge wird

$$\langle E''_m(0)\,h, h\rangle = H^2\frac{1-\mu}{4\pi_\mu}\int\limits_Q v_z^+ h\,\mathrm{d}x\,\mathrm{d}y\,. \tag{2.3}$$

Es ist besonders wichtig, daß sich die in die zweite Variation eingehenden Funktionen $v^\pm$ bei Kenntnis von $u^\pm$ durch ein weiteres quadratisches Variationsproblem kennzeichnen lassen. Zu seiner Herleitung differenzieren wir die Variationsgleichungen (1.12) nach t, wobei wir in (1.12) $\varphi^+ = \varphi|_{\Omega^+}$, $\varphi^- = \varphi|_{\Omega^-}$ mit einer den Festrandbedingungen (1.5), (1.8) genügenden Funktion $\varphi \in C^1(\bar\Omega)$ wählen. Man erhält für $t = 0$

$$\int\limits_{\Omega^+} \nabla v^+\,\nabla\varphi\,\mathrm{d}V - \int\limits_\Gamma \nabla u^+\,\nabla\varphi\,h\,\mathrm{d}x\,\mathrm{d}y + \mu\int\limits_{\Omega^-}\nabla v^-\,\nabla\varphi\,\mathrm{d}V + \mu\int\limits_\Gamma \nabla u^-\,\nabla\varphi\,h\,\mathrm{d}x\,\mathrm{d}y\,. \tag{2.4.}$$

Insbesondere finden wir für $\zeta = 0$

$$\int\limits_{\Omega^+} \nabla v^+\,\nabla\varphi\,\mathrm{d}V + \mu\int\limits_{\Omega^-}\nabla v^-\,\nabla\varphi\,\mathrm{d}V = 0$$

für die eben betrachteten Funktionen φ. Wir erschließen hieraus neben

$$\Delta v^+ = 0 \quad \text{in } \Omega^+\,, \qquad \Delta v^- = 0 \quad \text{in } \Omega^- \tag{2.5}$$

die Gültigkeit der Rand- bzw. Koppelungsbedingungen (1.9), (1.7) für die Funktionen v^+, v^-. Offenbar genügen diese Funktionen auch den Festrandbedingungen (1.5), (1.8) sowie der Bedingung (1.10). Sie sind daher Lösungen des Randwertproblems (1.4)—(1.10), wobei lediglich (1.6) durch (2.1) ($u_z^\pm = 0$) zu ersetzen ist.

Wir machen bei dieser Gelegenheit noch aufmerksam auf: Das in die allgemeineren Gleichungen (2.4) eingehende Randfunktional

$$l(\varphi) = -\int\limits_\Gamma h\nabla(u^+ - \mu u^-)\,\nabla\varphi\,\mathrm{d}x\,\mathrm{d}y$$

definiert bei $u^+, u^- \in C^{2+\alpha}$; $\zeta, h \in C^{2+\alpha}$ eine über dem SOBOLEWraum $V^+(V^-)$ stetige Linearform. Dies folgt aus (1.7) in Verbindung mit partieller Integration; es ist in leicht zu interpretierender Schreibweise

$$l(\varphi) = -\int\limits_\Gamma h\left(\frac{\partial}{\partial x}(u^+ - \mu u^-)\frac{\mathrm{d}\varphi}{\mathrm{d}x} + \frac{\partial}{\partial y}(u^+ - \mu u^-)\frac{\mathrm{d}\varphi}{\mathrm{d}y}\right)\mathrm{d}x\,\mathrm{d}y =$$

$$= \int\limits_\Gamma \left(\frac{\mathrm{d}}{\mathrm{d}x}h\frac{\partial}{\partial x}(u^+ - \mu u^-) + \frac{\mathrm{d}}{\mathrm{d}y}h\frac{\partial}{\partial y}(u^+ - \mu u^-)\right)\varphi\,\mathrm{d}x\,\mathrm{d}y\,.$$

Die Randintegrale entfallen wegen der Periodizität der beteiligten Funktionen. Dieser Sachverhalt gestattet es, (2.4) nachträglich ebenfalls als Variationsgleichungen über $V^+ \times V^-$ aufzufassen.

Wir studieren anschließend das Eigenwertproblem zur zweiten Variation $E''(0)$. Nach (1.2) ist

$$\langle E''(0)\,h_1, h_2\rangle = \beta\int\limits_Q \nabla h_1\,\nabla h_2\,\mathrm{d}x\,\mathrm{d}y + \varrho g\int\limits_Q h_1 h_2\,\mathrm{d}x\,\mathrm{d}y + \langle E''_m(0)\,h_1, h_2\rangle\,. \tag{2.6}$$

Bezeichnen wir die zu den Variationen h_i ($i = 1, 2$) gehörigen Ableitungen der Funktionen $u^\pm$ mit $v_i^\pm$, so gilt nach (2.3)

$$\langle E''_m(0)\,h_1, h_2\rangle = H^2\frac{1-\mu}{4\pi\mu}\int\limits_Q \frac{\partial v_1^+}{\partial z}\,h_2\,\mathrm{d}x\,\mathrm{d}y\,. \tag{2.7}$$

Insbesondere gilt wegen (2.1) ($u^+ = 0$, $u^- = 0$) und der oben bewiesenen Eigenschaften der Funktionen $v^\pm$

$$\langle E''_m(0)\, h, h\rangle = \frac{H^2}{4\pi} \int_Q \frac{\partial v^+}{\partial z} (v^+ - v^-)\, dx\, dy = -\frac{H^2}{4\pi} \int_{\Omega^+} |\nabla v^+|^2\, dV - \mu \frac{H^2}{4\pi} \int_{\Omega^-} |\nabla v^-|^2\, dV \leqq 0\,.$$

Berechnen wir zunächst das Spektrum des Operators $E''_m(0)$. Sei W der SOBOLEWraum der 2π-periodischen Funktionen $h(x, y)$ mit über Q quadratisch summierbaren ersten Ableitungen, außerdem gelte

$$\int_Q h(x, y)\, dx\, dy = 0$$

(vgl. (1.17)) für die Funktionen von W. Wegen $W \subset H^{1/2}(Q)$ und $v_z^\pm(x, y, 0) \in H^{-1/2}(Q)$ (s. [3]) läßt sich die Bilinearform (2.7) stetig über $W \times W$ fortsetzen. Entwickeln wir die Funktionen $h \in W$ in Fourierreihen

$$h(x, y) = \sum_{k\neq 0} \alpha_k\, e^{ikx}\,, \qquad \alpha_{-k} = \bar{\alpha}_k$$

— wir benutzen die Multiindexschreibweise: $k = (k_1, k_2)$ ganzzahlig, $|k|^2 = k_1^2 + k_2^2$, $kx = k_1 x + k_2 y$. Nach den sich an (2.5) anschließenden Überlegungen und (2.7) ist das Bildfunktional

$$E''_m(0)\, h = \sum_{k\neq 0} \alpha_k E''_m(0)\, e^{ikx}$$

wie folgt zu berechnen: Man löse zunächst das Randwertproblem (1.4)—(1.10), — (1.6) ist durch

$$v_k^+(x, y, 0) - v_k^-(x, y, 0) = \frac{1-\mu}{\mu}\, e^{ikx} + \text{const.}$$

zu ersetzen —, für die Funktionen $v_k^\pm$ und setze anschließend

$$E''_m(0)\, h = H^2 \frac{1-\mu}{4\pi\mu} \sum_{k\neq 0} \alpha_k \frac{\partial v_k^+}{\partial z} (x, y, 0)\,.$$

Man prüft leicht nach, daß

$$v_k^+ = A_k^+\, e^{ikx} \sinh |k|\, (z - z_1)\,, \qquad v_k^- = A_k^-\, e^{ikx} \sinh |k|\, (z + z_0)\,,$$

$$A_k^+ = (\mu - 1)\, (\mu \cosh |k|\, z_0 \sinh |k|\, z_1 + \cosh |k|\, z_1 \sinh |k|\, z_0)^{-1} \cosh |k|\, z_0\,,$$

$$A_k^- = \frac{\mu - 1}{\mu} (\mu \cosh |k|\, z_0 \sinh |k|\, z_1 + \cosh |k|\, z_1 \sinh |k|\, z_0)^{-1} \cosh |k|\, z_1\,,$$

also

$$\frac{\partial v_k^+}{\partial z} (x, y, 0) = \frac{(\mu - 1)\, |k|\, e^{ikx}}{\mu \tanh |k|\, z_1 + \tanh |k|\, z_0}$$

zutrifft. Gleichzeitig ist nachgewiesen, daß die Funktionen e^{ikx} Eigenfunktionen zu $E''_m(0)$ sind:

$$E''_m(0)\, e^{ikx} = H^2 \lambda_{|k|}^{(m)}\, e^{ikx}\,, \tag{2.8}$$

$$\lambda_{|k|}^{(m)} = -\frac{(1-\mu)^2\, |k|}{4\pi\mu\, (\mu \tanh |k|\, z_1 + \tanh |k|\, z_0)}\,.$$

Man beachte, daß die Höhen z_0, z_1 infolge unserer speziellen Periodenwahl als dimensionslos anzusehen sind.

Wir haben damit einen vollständigen Überblick über die quadratischen Anteile an der Magnetfeldenergie gewonnen. Fassen wir zusammen: Nach (2.8) gilt

$$\langle E''_m(0)\, h, h\rangle = 4\pi^2 H^2 \sum_{k\neq 0} \lambda_{|k|}^{(m)}\, |\alpha_k|^2 \geqq 4\pi^2 H^2 \lambda_1^{(m)} \sum_{k\neq 0} |k|\, |\alpha_k|^2 = H^2 \lambda_1^{(m)}\, ||h||_{1/2}^2; \tag{2.9.}$$

das Verhalten der quadratischen Terme (2.6) der Gesamtenergie wird demzufolge durch eine Ungleichung vom GÅRDINGschen Typ,

$$\langle E''(0), h, h\rangle \geqq \beta ||h||_1^2 + \varrho g ||h||_0^2 - H^2 |\lambda_1^{(m)}|\, ||h||_{1/2} \tag{2.10}$$

beschrieben. In (2.9), (2.10) sind

$$||h||_1^2 = \int_Q |\nabla h|^2\, dx\, dy\,, \qquad ||h||_0^2 = \int_Q |h|^2\, dx\, dy\,,$$

$$||h||_{1/2}^2 = 4\pi^2 \sum_{k\neq 0} |k|\, |\alpha_k|^2$$

die Normen in W, $L_2(Q)$ und $H^{1/2}(Q)/R$. Da die Funktionen e^{ikx} auch Eigenfunktionen des Kapillar-Schwereanteils der Energie sind — und zwar zu den Eigenwerten $\beta\, |k|^2 + \varphi g$ —, gilt

$$\langle E''(0)\, h, h\rangle = 4\pi^2 \sum_{k\neq 0} (\beta\, |k|^2 + \varphi g + H^2 \lambda_{|k|}^{(m)})\, |\alpha_k|^2\,.$$

Die horizontale Trennfläche $\zeta = 0$ verhält sich deshalb in unserer Lösungsklasse stabil, solange $\beta\,|k^2| + \varphi g + H^2 \lambda_{|k|}^{(m)} > 0$ zutrifft, d. h. die Magnetfeldstärke in dem hierdurch bestimmten unterkritischen Bereich verbleibt.

Literatur

1 Gailitis, A., Formation of the hexagonal pattern on the surface of a ferromagnetic fluid in an applied magnetic field, J. Fluid Mech. 82, 401—413 (1977).
2 Landau, L. D., Lifschitz, E. M., Elektrodynamik der Kontinua, Berlin 1967.
3 Lions, J.-L., Magenes, E., Problèmes aux limites non homogènes et applications I, Paris 1968.
4 Whitham, G. B., Linear and Nonlinear Waves, New York 1974.
5 Zeidler, E., Zur Verzweigungstheorie und zur Stabilitätstheorie der Navier-Stokesschen Gleichungen, Math. Nachr. 52, 167—205 (1972).

Eingereicht am 8. 10. 1979

Anschrift: Prof. Dr. Klaus Beyer, Sektion Mathematik, Wilhelm-Pieck-Universität Rostock, 2500 Rostock, Universitätsplatz, DDR

ZAMM 60, 241 – 248 (1980)

WOON-SHING YEUNG

Perturbation Solutions for the Particle Trajectories of a Gas-Solid Mixture Entering a Curved Duct

A matched asymptotic expansion solution was found for the particle trajectories in a curved duct when the non-dimensional momentum equilibration length $\tilde{L}_m$ is small. The motion of the carrier fluid was assumed to be uniform. The approximate solution was compared with the exact numerical solution and good agreements were found.

Für den Fall kleiner dimensionsloser Impulsgleichgewichtslänge $\tilde{L}_m$ wurde für die Partikeltrajektorien in einem gekrümmten Rohr eine asymptotische Lösung gefunden. Die Bewegung des Trägerfluids wurde als gleichförmig angenommen. Die Näherungslösung wurde mit der exakten numerischen Lösung verglichen. Es wurde gute Übereinstimmung festgestellt.

В случае малой безразмерной длины уравновешенного импульса $\tilde{L}_m$ было найдено для траекторий частичек в изогнутой трубе подходящее асимптотическое решение. Движение жидкости-носителя принималось равномерным. Приближённое решение сравнивалось с точным численным решением причём было установлено хорошее совпадение.

Introduction

In a previous study, YEUNG [1] has obtained numerical solutions for the particle trajectories in a circular curved tube. Two dimensionless quantities were identified: $\tilde{L}_m$, the nondimensional momentum equilibration length, and δ, the ratio of the mean radius of the elbow to the radius of the cross section of the pipe. It is thus natural to investigate the perturbation solution when $\tilde{L}_m$ or δ, or both, are small (or large). The main advantage of perturbation solutions is that the variables of the governing equations can be represented approximately in closed forms. These are considerably better than the numerical solutions. For example, the calculation of particle density in the region of interest can be more easily carried out using the approximate closed form solution than the numerical solution. It also saves a lot of computer storage spaces when momentum coupling between the gas and the particle phase is accounted for.

Theory

Consider a gas-particle mixture flowing into a 2-D curve dduct with mean radius of curvature R as shown in Fig. 1.

A 2-D curved duct is chosen for simplification, so that δ is absent from the governing equations. The main assumptions are those used in the previous study [1], which are listed below:

(1) Particle-particle interaction is small compared to particle-fluid interaction.
(2) The presence of particles does not influence the gas flow field.
(3) Only the drag force exerted by the fluid phase on the particle phase is considered.
(4) The aerodynamic drag force is assumed to be given by STOKES law throughout the region of interest.
(5) The fluid moves with uniform speed along the curved pipe.

Some justifications of the above assumptions can be found in the previous investigation [1]. If we denote u as the velocity component in increasing r-direction and v as the velocity component in the increasing ψ-direction, we can write down the motion of the fluid phase in view of assumption (5):

$$u_f = 0 , \tag{1}$$

$$v_f = W , \tag{2}$$

where W is the initial entry velocity of the dilute suspension and subscript f denotes values pertaining to the fluid phase. For the motion of the particle phase, we shall use the Lagrangian description. As mentioned by YEUNG [2], collisions among particles cannot be neglected if one tries to formulate the motion of an individual particle, which presents great difficulty in the solution process. However, when the particle phase is considered as a continuum, the particle-particle interaction becomes negligible compared with the particle-fluid interaction, as long as the

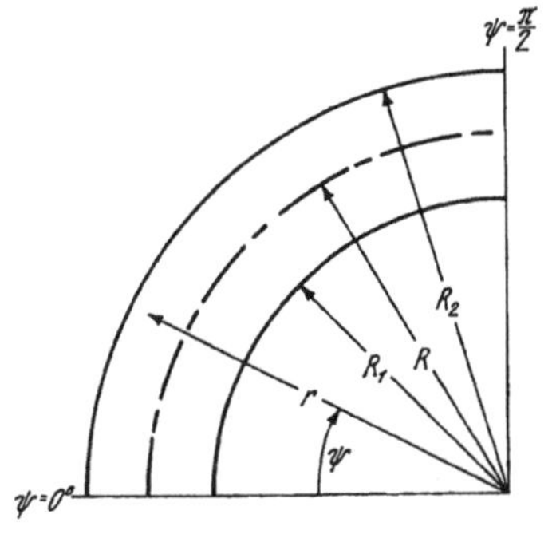

Fig. 1. Coordinate system of the geometry used

suspension is dilute [2]. Thus, if the particle phase behaves as, or at least approximately, a continuum, the momentum equations of the particle phase can be expressed as

$$\frac{\partial u_p}{\partial t} + u_p \frac{\partial u_p}{\partial r} + \frac{v_p}{r} \frac{\partial u_p}{\partial \psi} - \frac{v_p^2}{r} = \frac{u_f - u_p}{\tau_m}, \tag{3}$$

$$\frac{\partial v_p}{\partial t} + u_p \frac{\partial v_p}{\partial r} + \frac{v_p}{r} \frac{\partial v_p}{\partial \psi} + \frac{u_p v_p}{r} = \frac{v_f - v_p}{\tau_m}, \tag{4}$$

where we have neglected the diffusional stress tensor of the particle phase due to its interaction with the fluid phase. τ_m is the momentum equilibration time and the subscript p denotes the particle phase. By virtue of assumption (4), τ_m is given by

$$\tau_m = \frac{2}{9} \frac{\sigma^2 \bar{\varrho}_p}{\mu_f} \tag{5}$$

where σ denotes the particle diameter, $\bar{\varrho}_p$ the particle material density and μ_f the viscosity of the fluid phase. It is of particular importance to realize that equations (3) and (4) describe the mean motion of the particulate phase, but not the motion of individual particles. Thus, imagine a small volume surrounding a point in space, which is large enough to contain a great number of particles, while still possessing dimensions small compared with the characteristic dimension of the physical system. The quantities u_p, v_p in equations (3) and (4) then represent the average value of the motion of all the particles contained in that small volume. Since only mean values are of main concern in most practical situations, the Lagrangian formulation of individual particle motion appears to be more than adequate. Moreover, the collisions among particles render the Lagrangian formulation impractical. To circumvent this difficulty, the notion of emsemble average*) is used with the Lagrangian formulation of the particle motion. To begin with, we write down the equation of motion of a particle

$$M \frac{d\boldsymbol{v}}{dt} = F(t), \tag{6}$$

where M is the particle mass, $\boldsymbol{v}$ the particle velocity and $F(t)$ the force acting upon the particle by virtue of collisions from the fluid molecules and other particles. Following LANGEVIN [3] on the theory of Brownian motion, we assume $F(t)$ can be written as a sum of two parts: (i) an "averaged-out" part, which represents the viscous drag experienced by the particle and (ii) a "fluctuating" part $f(t)$. Equation (6) becomes

$$M \frac{d\boldsymbol{u}_p}{dt} = M \frac{\boldsymbol{v}_f - \boldsymbol{v}_p}{\tau_m} + f(t). \tag{7}$$

We now consider an ensemble of a large number of systems and take the ensemble average of (7),

$$M \frac{d\langle \boldsymbol{u}_p \rangle}{dt} = M \frac{\langle \boldsymbol{v}_f \rangle - \langle \boldsymbol{u}_p \rangle}{\tau_m} + \langle f(t) \rangle. \tag{8}$$

The important assumption to be made here is

$$\langle f(t) \rangle = 0 \tag{9}$$

and equation (8) is reduced to

$$\frac{d\langle \boldsymbol{v}_p \rangle}{dt} = \frac{\langle \boldsymbol{v}_f \rangle - \langle \boldsymbol{v}_p \rangle}{\tau_m}. \tag{10}$$

The particle may collide with other particles and thereby loses momentum or changes direction. This may be looked upon as a drag force acting on the particle by all other particles. However, since the collision frequency of the particle with the gas molecules is much larger than with the other particles, the averaged drag force is represented solely by the viscous drag due to the fluid. Equation (9) is justified by the random nature of $f(t)$**). Equation (10) is the necessaryequation of motion of a particle in terms of the ensemble average value of the velocity. Since the ensemble average value is the one most likely to be observed macroscopically, it can be looked upon as the mean motion of the particle phase. From here on, we drop the use of brackets in equation (10). It should be clear that only mean values are formulated and solved for. Hence

$$\frac{d\boldsymbol{v}_p}{dt} = \frac{\boldsymbol{v}_f - \boldsymbol{v}_p}{\tau_m}. \tag{11}$$

In polar coordinates,

$$\boldsymbol{v}_p = u_p \boldsymbol{e}_r + v_p \boldsymbol{e}_\psi, \tag{12}$$

*) An ensemble is a collection of a large number of systems similar to the one originally under consideration. The ensemble average is one which is taken over the ensemble at any instant of time. It was introduced by GIBBS and is used extensively in statistical mechanics.

**) Strictly speaking, a special proof is required. It may be considered as a pure assumption in the present work.

where e_r and e_ψ are the unit vectors in the increasing r and ψ directions, respectively. Substituting (12), (1) and (2) into equation (11) and remembering that e_r and e_ψ are not constant vectors, we obtain

$$\frac{du_p}{dt} - \frac{v_p^2}{r} = -\frac{u_p}{\tau_m} \tag{13}$$

in the r-direction and

$$\frac{dv_p}{dt} + \frac{u_p v_p}{r} = \frac{W - v_p}{\tau_m} \tag{14}$$

in the ψ direction. We also have

$$\frac{dr}{dt} = u_p \tag{15}$$

and

$$\frac{d\psi}{dt} = \frac{v_p}{r}. \tag{16}$$

Equations (13) to (16) can be solved for the particle trajectories under some prescribed initial conditions. One may have obtained the same set of governing equations by considering the motion of an individual particle under the influence of only the viscous drag of the fluid phase. In that case, $f(t)$ would have been identically zero at any instant, which is a highly improbable situation even when there is absolutely no collision among particles*). Indeed, the autocorrelation function of $f(t)$ is important in some other quantities of interest in the theory of Brownian motion [4]. Thus, the derivation given above of equations (13) to (16) is fundamentally more correct and rigorous. Furthermore, the use of ensemble average values of macroscopic quantities is more appropriate for most physical systems in practice.

For the initial conditions, we assume the gas and praticle phases are in dynamic equilibrium and that the particles are uniformly distributed at the entry. Hence the ensembled averages of the initial conditions are

$$u_p = 0\,, \qquad v_p = W\,, \qquad r = r_0\,, \qquad \psi = 0 \tag{17}$$

at $t = 0$, and $(r_0, 0)$ being the initial position of the particle.

Matched Asymptotic Solution

As mentioned before, we are seeking analytical solutions for the particle trajectories when $\tilde{L}_m$, the nondimensional momentum equilibration length, is small. It can be readily shown, by nondimensionalizing the particle equations, that $\tilde{L}_m$ is given by

$$\tilde{L}_m = \frac{W\tau_m}{R} = \frac{\tau_m}{R/W} = \frac{\tau_m}{t^*}, \tag{18}$$

where

$$t^* = R/W\,. \tag{19}$$

Hence small $\tilde{L}_m$ implies that

$$t^* \gg \tau_m\,. \tag{20}$$

It is well known in perturbation theory that if the small parameter, in our case $\tilde{L}_m$, is a ratio of two time scales, in our case τ_m and t^*, it will be a singular perturbation problem [5]. One cannot generate series solutions, which are uniformly valid in the domain of interest, by straightforward expansion. Several methods exist to deal with singular perturbation problems and the method of matched asymptotic expansions is most suitable for the present problem.

We begin by identifying the outer variable and the inner variable of the problem, which are t^* and τ_m, respectively. Then the outer problem is formulated by introducing the following dimensionless quantities:

$$\bar{U} = \frac{u_p}{W}, \qquad \bar{V} = \frac{v_p}{W}, \qquad \bar{R} = \frac{r}{R}, \qquad \bar{\psi} = \psi\,, \qquad \bar{t} = \frac{t}{t^*}. \tag{21}$$

Equations (13) to (16) become

$$\tilde{L}_m \frac{d\bar{U}}{d\bar{t}} = -\bar{U} + \tilde{L}_m \frac{\bar{V}^2}{\bar{R}}, \qquad \tilde{L}_m \frac{d\bar{V}}{d\bar{t}} = 1 - \bar{V} - \tilde{L}_m \frac{\bar{U}\bar{V}}{\bar{R}}, \tag{22), (23}$$

$$\frac{d\bar{R}}{d\bar{t}} = \bar{U}\,, \qquad \frac{d\bar{\psi}}{d\bar{t}} = \frac{\bar{V}}{\bar{R}}. \tag{24), (25}$$

Examination of equations (22) and (23) shows that when $\tilde{L}_m \to 0$, they become algebraic equations instead of first order ordinary differential equations. The given initial conditions (17) cannot be satisfied consequently. The failure

*) Recall that $f(t)$ is the fluctuating part of the force due to particle-molecule and particle-particle collisions as well.

to satisfy all the prescribed boundary and initial conditions of the outer solution is typical in all singular perturbation problems. Hence, instead of using (17), one must obtain the general solutions of (23) to (25) and evaluate any constants in the general solution by matching with the inner solution. Pertinent solutions of (22) to (25) are written in the forms

$$\bar{U} = \bar{U}_0 + \tilde{L}_m\bar{U}_1 + \tilde{L}_m^2\bar{U}_2 + \dots, \qquad \bar{V} = \bar{V}_0 + \tilde{L}_m\bar{V}_1 + \tilde{L}_m^2\bar{V}_2 + \dots, \tag{26), (27}$$

$$\bar{R} = \bar{R}_0 + \tilde{L}_m\bar{R}_1 + \tilde{L}_m\bar{R}_m^2 + \dots, \qquad \bar{\psi} = \bar{\psi}_0 + \tilde{L}_m\bar{\psi}_1 + \tilde{L}_m^2\bar{\psi}_2 + \dots. \tag{28), (29}$$

Substituting (26), (27); (28), (29) into equations (22) to (25) and equating equal order of $\tilde{L}_m$ we obtain

$$0 = -\bar{U}_0\,, \qquad 0 = 1 - \bar{V}_0\,, \qquad \dot{\bar{R}}_0 = \bar{U}_0\,, \qquad \dot{\bar{\psi}}_0 = \frac{\bar{V}_0}{\bar{R}_0} \tag{30}$$

for the zeroth order and

$$\dot{\bar{U}}_0 = -\bar{U}_1 + \bar{V}_0^2/\bar{R}_0\,, \qquad \dot{\bar{V}}_0 = -\bar{V}_1 - \frac{\bar{V}_0\bar{U}_0}{\bar{R}_0}, \qquad \dot{\bar{R}}_1 = \bar{U}_1\,, \qquad \dot{\bar{\psi}}_1 = \frac{1}{\bar{R}_0}\left(\bar{V}_1 - \frac{\bar{R}_1}{\bar{R}_0}\bar{V}_0\right) \tag{31}$$

for the first order of $\tilde{L}_m$. The dot denotes differentiation with respect to $\bar{t}$. The governing equations for higher order terms are complicated algebraically and are not recorded here.

The general solution to the outer problem is found, to order of $\tilde{L}_m^3$, as follows:

$$\bar{U} = \frac{1}{A}\tilde{L}_m - \frac{1}{A^2}\left(\frac{\bar{t}}{A} + C\right)\tilde{L}_m^2 + \frac{1}{A}\left(-\frac{1}{A^2} - \frac{E}{A} + \frac{C^2}{A^2} + \frac{3C}{A^2}\bar{t} + \frac{3}{2A^4}\bar{t}^2\right)\tilde{L}_m^3 + O(\tilde{L}_m^4)\,, \tag{32)*}$$

$$\bar{V} = 1 - \frac{1}{A^2}\tilde{L}_m^2 + \frac{1}{A}\left[\frac{2}{A^2}\left(\frac{\bar{t}}{A} + C\right)\right]\tilde{L}_m^3 + O(\tilde{L}_m^4)\,, \tag{33}$$

$$\bar{R} = A + \left(C + \frac{\bar{t}}{A}\right)\tilde{L}_m + \left[E - \frac{1}{A^2}\left(C\bar{t} + \frac{\bar{t}^2}{2A}\right)\right]\tilde{L}_m^2 + $$
$$+ \left\{\frac{1}{A}\left[\bar{t}\left(-\frac{1}{A^2} - \frac{E}{A} + \frac{C^2}{A^2}\right) + \frac{3C}{2A^3}\bar{t}^2 + \frac{1}{2A^4}\bar{t}^3\right] + G\right\}\tilde{L}_m^3 + O(\tilde{L}_m^4)\,, \tag{34}$$

$$\bar{\psi} = \left(\frac{\bar{t}}{A} + B\right) + \left[D - \frac{1}{A^2}\left(C\bar{t} + \frac{\bar{t}^2}{2A}\right)\right]\tilde{L}_m + \left\{F + \frac{1}{A}\left[\bar{t}\left(\frac{-1}{A^2} - \frac{E}{A}\right) + \frac{C}{2A^3}\bar{t}^2 + \frac{1}{6A^4}\bar{t}^3\right]\right\}\tilde{L}_m^2 + $$
$$+ \left\{\frac{1}{A}\left[\left(\frac{3C}{A^3} + \frac{2EC}{A^2} - \frac{G}{A} - \frac{C^3}{A^3}\right)\bar{t} + \frac{\bar{t}^2}{2}\left(\frac{4}{A^4} + \frac{3E}{A^3} - \frac{6C^2}{A^4}\right) + \frac{\bar{t}^3}{3}\left(-\frac{15}{2}\frac{C}{A^3}\right) + \frac{\bar{t}^4}{4}\left(-\frac{5}{2A^6}\right)\right] + H\right\}\tilde{L}_m^3\,, \tag{35}$$

where A, B, C, D, E, F, G, H are constants to be determined by matching conditions.

For the inner problem, the following nondimensional quantities are used:

$$U = \frac{u}{W}, \qquad V = \frac{v}{W}, \qquad R = \frac{r}{R}, \qquad \Psi = \psi\,, \qquad T = \frac{t}{\tau_m}, \tag{36}$$

where we have nondimensionalized t with the inner variable τ_m instead of t^*. T and $\bar{t}$ are related by the following

$$T = \frac{\bar{t}}{\tilde{L}_m}. \tag{37}$$

Substitute (36) into equations (13) to (16) to obtain

$$\frac{dU}{dT} = -U + \tilde{L}_m\frac{V^2}{R}; \qquad \frac{dV}{dT} = 1 - V - \frac{VU}{R}\tilde{L}_m\,, \tag{38), (39}$$

$$\frac{dR}{dT} = U\tilde{L}_m\,, \qquad \frac{d\Psi}{dT} = \frac{V}{R}\tilde{L}_m. \tag{40, (41}$$

The main difference between the outer and the inner problem is that the latter does not alter the order of the original system of differential equations, as seen from (38) to (41). Thus, the initial conditions (17) can be satisfied by the inner solution. As before, U, V, R and Ψ are assumed to be of the following forms

$$U = U_0 + \tilde{L}_mU_1 + \tilde{L}_m^2U_2 + \dots, \qquad V = V_0 + \tilde{L}_mV_1 + \tilde{L}_m^2V_2 + \dots, \tag{42), (43}$$

$$R = R_0 + \tilde{L}_mR_1 + \tilde{L}_m^2R_2 + \dots, \qquad \Psi = \Psi_0 + \tilde{L}_m\Psi_1 + \tilde{L}_m^2\Psi_2 + \dots. \tag{44), (45}$$

The initial conditions (17) become

$$U_0 = 0\,, \quad V_0 = 1\,, \quad R_0 = \bar{r}_0\,, \quad \Psi_0 = 0 \quad \text{and} \quad U_i = 0, \quad V_i = 0, \quad R_i = 0, \quad \Psi_i = 0; \quad \forall i \geqq 1 \quad \text{at} \quad T = 0, \tag{46}$$

*) In view of equation (32), u_p should have been nondimensionalized according to $\bar{U} = u_p/W\tilde{L}_m$ so as to render $\bar{U}$ of order 1. However, the final results were found to be identical with the present results when the new form of $\bar{U}$ was used.

where

$$\bar{r}_0 = \frac{r_0}{R}. \tag{47}$$

Substituting equations (42) to (45) into (38) to (41) and equating equal orders of $\tilde{L}_m$ as before, we obtain systems of differential equations, subject to the appropriate initial conditions given by (46). The solution procedure is straightforward and the inner expansions of U, V, R and Ψ are found to be, to order $O(\tilde{L}_m^5)$, the following:

$$U = \frac{1}{\bar{r}_0}(1 - e^{-T})\,\tilde{L}_m + \frac{1}{\bar{r}_0^3}(T e^{-T} + T^2 e^{-T} - T)\,\tilde{L}_m^3 + \\ + \frac{1}{\bar{r}_0^5}\left(\frac{3}{2}T^2 + 8e^{-T} + 13T e^{-T} + \frac{3T^2}{2}e^{-T} - \frac{4T^3}{3}e^{-T} - \frac{1}{6}T^4 e^{-T} + T - 8 - 2T e^{-2T}\right)\tilde{L}_m^5 + O(\tilde{L}_m^6), \tag{48}$$

$$V = 1 + \frac{1}{\bar{r}_0^2}(e^{-T} + T e^{-T} - 1)\,\tilde{L}_m^2 + \frac{1}{\bar{r}_0^4}\left(2T - 2 + 3e^{-T} - \frac{3}{2}T^2 e^{-T} - \frac{T^3}{3}e^{-T} - T e^{-2T} - e^{-2T}\right)\tilde{L}_m^4 + \\ + O(\tilde{L}_m^6), \tag{49}$$

$$R = \bar{r}_0 + \frac{1}{\bar{r}_0}(T + e^{-T} - 1)\,\tilde{L}_m^2 + \frac{1}{\bar{r}_0^3}\left(3 - 3e^{-T} - 3T e^{-T} - \frac{T^2}{2} - T^2 e^{-2T}\right)\tilde{L}_m^4 + O(\tilde{L}_m^6), \tag{50}$$

$$\Psi = \frac{T}{\bar{r}_0}\tilde{L}_m + \frac{1}{\bar{r}_0^3}\left(1 - e^{-T} - T e^{-T} - \frac{T^2}{2}\right)\tilde{L}_m^3 + \\ + \frac{1}{\bar{r}_0^5}\left\{9 - 5T + \frac{T^2}{2} + \frac{T^3}{6} - 10e^{-T} + \frac{5}{2}T^2 e^{-T} + \frac{1}{3}T^3 e^{-T} + T e^{-T} + e^{-2T}\right\}\tilde{L}_m^5 + O(\tilde{L}_m^6). \tag{51}$$

Applying VAN DYKE's matching principle [5], the constants appear in the outer solution (32) to (35) are found as

$$A = \bar{r}_0, \quad B = 0, \quad C = 0, \quad D = 0, \quad E = -1/\bar{r}_0, \quad F = 0, \quad G = 0, \quad H = \frac{1}{3}. \tag{52}$$

The outer solution then becomes

$$\bar{U} = \frac{1}{\bar{r}_0}\tilde{L}_m - \frac{\bar{t}}{\bar{r}_0^3}\tilde{L}_m^2 + \frac{3}{2\bar{r}_0^5}\bar{t}^2\tilde{L}_m^3 + O(\tilde{L}_m^4), \tag{53}$$

$$\bar{V} = 1 - \frac{1}{\bar{r}_0^2}\tilde{L}_m^2 + \frac{2\bar{t}}{\bar{r}_0^4}\tilde{L}_m^3 + O(\tilde{L}_m^4), \tag{54}$$

$$\bar{R} = \bar{r}_0 + \frac{\bar{t}}{\bar{r}_0}\tilde{L}_m + \left(-\frac{1}{\bar{r}_0} - \frac{\bar{t}^2}{2\bar{r}_0^3}\right)\tilde{L}_m^2 + \left(\frac{\bar{t}^3}{2\bar{r}_0^5}\right)\tilde{L}_m^3 + O(\tilde{L}_m^4), \tag{55}$$

$$\bar{\Psi} = \frac{\bar{t}}{\bar{r}_0} - \frac{\bar{t}^2}{2\bar{r}_0^3}\tilde{L}_m + \frac{\bar{t}^3}{6\bar{r}_0^5}\tilde{L}_m^2 + \left(\frac{\bar{t}^2}{2\bar{r}_0^5} - \frac{5}{8}\frac{\bar{t}^4}{\bar{r}_0^7} + \frac{1}{\bar{r}_0^3}\right)\tilde{L}_m^3 + O(\tilde{L}_m^4). \tag{56}$$

Equations (48) to (51) and equations (53) to (56) then constitute the solution of the particle motion governed by equations (13) to (16) when $\tilde{L}_m$ is small. For time of order τ_m, the inner solution applies and for time of order t^*, the outer solution applies. However, it is quite inconvenient to use two separate expansions from a practical viewpoint and a composite expansion valid for all time is desirable. One such composite expansion is given by ERDELYI [6]

$$f_c^{(\delta,\Delta)} = f_i^{(\delta)} + f_0^{(\Delta)} - [f_0^{(\Delta)}]_i^{(\delta)}, \tag{57}$$

where $f_c^{(\delta,\Delta)}$ is the composite expansion, $f_i^{(\delta)}$ is the inner expansion to order δ, $f_0^{(\Delta)}$ is the outer expansion to order Δ, and $[f_0^{(\Delta)}]_i^{(\delta)}$ denotes the inner expansion to order δ of the outer expansion $f_0^{(\Delta)}$.

Using (57), the composite expansions for U, V, R and Ψ are

$$U = \frac{L_m}{\bar{r}_0} - \frac{\bar{t}}{\bar{r}_0^3}\tilde{L}_m^2 + \frac{3}{2}\frac{\bar{t}^2}{\bar{r}_0^5}\tilde{L}_m^3 + \frac{\bar{t}}{\bar{r}_0^5}\tilde{L}_m^4 - \frac{8}{\bar{r}_0^5}\tilde{L}_m^5 + \\ + e^{-\bar{t}/\tilde{L}_m}\left\{\left(-\frac{1}{\bar{r}_0} + \frac{\bar{t}^2}{\bar{r}_0^3} - \frac{1}{6}\frac{\bar{t}^4}{\bar{r}_0^5}\right)\tilde{L}_m + \left(\frac{\bar{t}}{\bar{r}_0^3} - \frac{4}{3}\frac{\bar{t}^3}{\bar{r}_0^5}\right)\tilde{L}_m^2 + \left(\frac{3}{2}\frac{\bar{t}^2}{\bar{r}_0^5}\right)\tilde{L}_m^3 + \frac{13\bar{t}}{\bar{r}_0^5}\tilde{L}_m^4 + \frac{8}{\bar{r}_0^5}\tilde{L}_m^5\right\} + \\ + e^{-2\bar{t}/\tilde{L}_m}\left(-\frac{2\bar{t}}{\bar{r}_0^5}\tilde{L}_m^4\right), \tag{58}$$

$$V = 1 - \frac{1}{\bar{r}_0^2}\tilde{L}_m^2 + \frac{2\bar{t}}{\bar{r}_0^4}\tilde{L}_m^3 - \frac{2}{\bar{r}_0^4}\tilde{L}_m^4 + \\ + e^{-\bar{t}/\tilde{L}_m}\left\{\left(\frac{\bar{t}}{\bar{r}_0^2} - \frac{\bar{t}^3}{3\bar{r}_0^4}\right)\tilde{L}_m + \left(\frac{1}{\bar{r}_0^4} - \frac{3}{2}\frac{\bar{t}^2}{\bar{r}_0^4}\right)\tilde{L}_m^2 + \left(\frac{3}{\bar{r}_0^4}\right)\tilde{L}_m^4\right\} + e^{-2\bar{t}/\tilde{L}_m}\left\{-\frac{\bar{t}}{\bar{r}_0^4}\tilde{L}_m^3 - \frac{1}{\bar{r}_0^4}\tilde{L}_m^4\right\}, \tag{59}$$

$$R = \bar{r}_0 + \frac{\bar{t}}{\bar{r}_0}\tilde{L}_m + \left(-\frac{1}{\bar{r}_0} - \frac{\bar{t}^2}{2\bar{r}_0^3}\right)\tilde{L}_m^2 + \frac{1}{2}\frac{\bar{t}^3}{\bar{r}_0^5}\tilde{L}_m^3 + \frac{3}{\bar{r}_0^3}\tilde{L}_m^4 + + e^{-\bar{t}/\tilde{L}_m}\left\{\left(\frac{1}{\bar{r}_0} - \frac{\bar{t}^2}{\bar{r}_0^3}\right)\tilde{L}_m^2 - \frac{3\bar{t}}{\bar{r}_0^3}\tilde{L}_m^3 - \frac{3}{\bar{r}_0^3}\tilde{L}_m^4\right\}, \tag{60}$$

$$\Psi = \frac{\bar{t}}{\bar{r}_0} - \frac{\bar{t}^2}{2\bar{r}_0^3}\tilde{L}_m + \frac{\bar{t}^3}{6\bar{r}_0^5}\tilde{L}_m^2 + \left(\frac{1}{\bar{r}_0^3} + \frac{\bar{t}^2}{2\bar{r}_0^5} - \frac{5}{8}\frac{\bar{t}^4}{\bar{r}_0^7}\right)\tilde{L}_m^3 - \frac{5\bar{t}}{\bar{r}_0^5}\tilde{L}_m^4 + \frac{9}{\bar{r}_0^5}\tilde{L}_m^5 + + e^{-\bar{t}/\tilde{L}_m}\left\{\left(-\frac{\bar{t}}{\bar{r}_0^3} + \frac{\bar{t}^3}{3\bar{r}_0^5}\right)\tilde{L}_m^2 + \left(-\frac{1}{\bar{r}_0^3} + \frac{5}{2}\frac{\bar{t}^2}{\bar{r}_0^5}\right)\tilde{L}_m^3 - \frac{10}{\bar{r}_0^5}\tilde{L}_m^5\right\} + e^{-2\bar{t}/\tilde{L}_m}\left\{\frac{\bar{t}}{\bar{r}_0^5}\tilde{L}_m^4 + \frac{1}{\bar{r}_0^5}\tilde{L}_m^5\right\}. \tag{61}$$

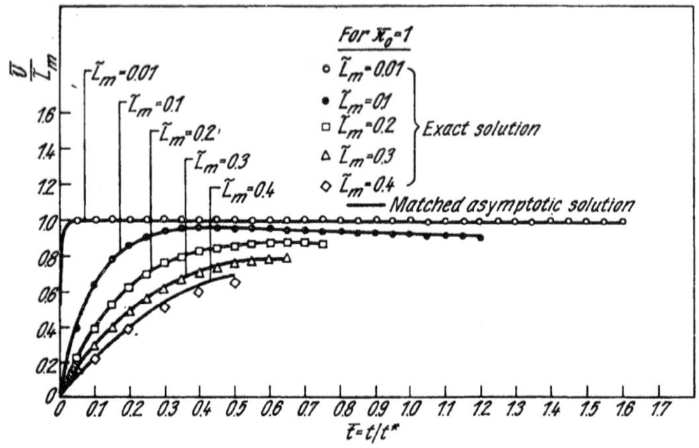

Fig. 2. Comparison of the approximate solution to the exact numerical solution of the radial velocity component

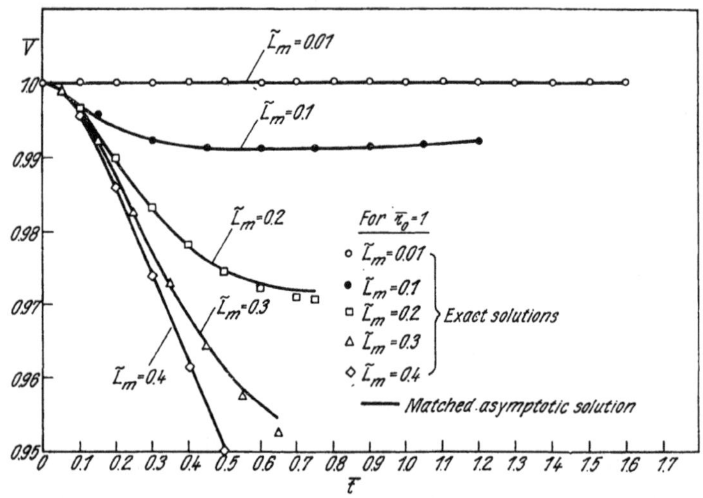

Fig. 3. Comparison of the approximate solution to the exact numerical solution of the circumferential velocity component

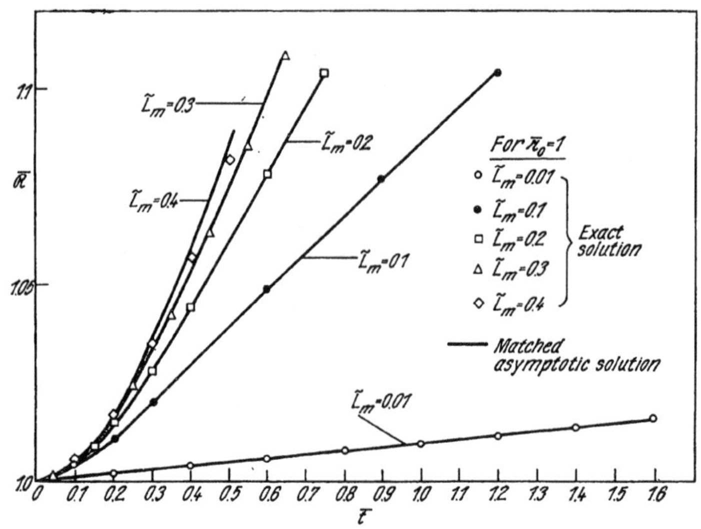

Fig. 4. Comparison of the approximate solution to the exact numerical solution of the radial coordinate

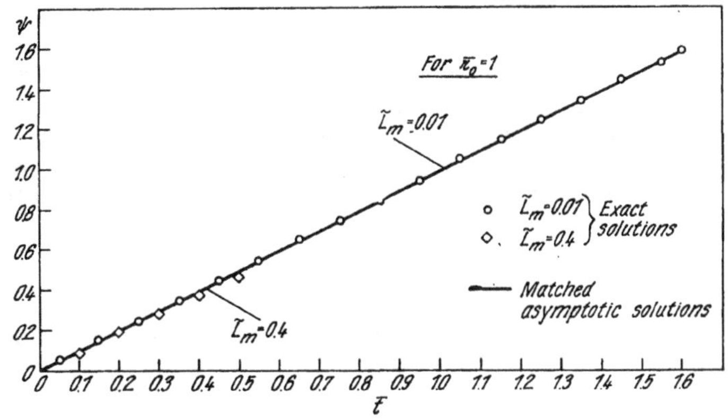

Fig. 5. Comparison of the approximate solution to the exact numerical solution of the angular coordinate

Results and Discussion

The exact solutions can be found by numerically integrating equations (22) to (25), subject to the initial conditions (17) nondimensionalized as follows

$$\text{at}\quad \bar t = 0\,,\quad \bar U = 0\,,\quad \bar V = 1\,,\quad \bar R = r_0\,,\quad \bar\Psi = 0\,. \tag{62}$$

Figures 2 to 5 show the comparison between the exact numerical solution and the approximate solution given by the composite expansions (58) to (61) for various values of $\tilde L_m$. Excellent agreements between the numerical exact solutions and the matched asymptotic solutions are indicated for $\tilde L_m$ as large as 0.4. For large values of $\tilde L_m$ (but still less than 1), the approximate solutions tend to over predict the exact values in general. The comparisons are made for particle trajectories initiated at $\bar r_0 = 1$, i.e., center of the curve duct. Also, the results are presented up to the time when the particle hits the pipe wall or exits the pipe. Having obtained the approximate solution for the particle motion, one can calculate, for example, the erosion rate on the pipe wall due to particle impacts [1].

The present results also apply to the mid-plane of a circular curved pipe, as long as the fluid motion is uniform over the cross-section and along the pipe [1]. To show this analytically, we notice that the particle velocity is

$$\boldsymbol v_p = u_p \boldsymbol e_r + v_p \boldsymbol e_\psi + w_p \boldsymbol e_z \tag{63}$$

written in polar coordinates with the z-axis pointing into the paper in Fig. 1. w_p is the velocity component in the increasing z-direction. Carrying out the same analysis as before, we arrive at the following governing equation for the ensemble average of the velocity components:

$$\frac{du_p}{dt} - \frac{v_p^2}{r} = -\frac{u_p}{\tau_m}\,,\qquad \frac{dv_p}{dt} + \frac{u_p v_p}{r} = \frac{W - v_p}{\tau_m}\,,\qquad \frac{dw_p}{dt} = -\frac{w_p}{\tau_m}\,. \tag{64), (65), (66}$$

$$\frac{du}{dt} = u_p\,,\qquad \frac{d\psi}{dt} = \frac{v_p}{r}\,,\qquad \frac{dz}{dt} = w_p\,, \tag{67), (68), (69}$$

with initial conditions

$$u_p = 0\,,\quad v_p = W\,,\quad w_p = 0\,,\quad r = r_0\,,\quad \psi = 0\,,\quad z = z_0 \tag{70}$$

at $t = 0$. The initial position of the particle is now $(r_0, 0, z_0)$. The general solution to (66) is

$$w_p = \text{const.}\ e^{-t/\tau_m}\,, \tag{71}$$

which gives, by virtue of the initial condition on w_p,

$$w_p = 0\,,\quad \forall t\,. \tag{72}$$

Hence equation (69) becomes

$$z = \bar z_0\,,\quad \forall t\,. \tag{73}$$

The remaining equations (64), (65), (67) and (68) and the initial conditions are seen to be identical with equations (13)—(16) and (17). Hence equations (58) to (61) together with (72) and (73) constitute the matched asymptotic solutions to equations (64)—(69). The particle trajectory will hit the pipe wall when the following is satisfied

$$z^2 + (R - r)^2 = a^2\,, \tag{74}$$

where R is the mean radius of the elbow as before and a is the radius of the circular cross-section in this case. By virtue of (73), (74) can be solved for r to obtain

$$r = R - \sqrt{a^2 - \bar z_0^2}\quad \text{for}\quad R > r \tag{75}$$

and

$$r = R + \sqrt{a^2 - \bar z_0^2}\quad \text{for}\quad R < r \tag{76}$$

as the conditions that the particle hits the wall.

Conclusion

The method of matched asymptotic expansion has been applied successfully to obtain series solutions for the particle trajectories when $\tilde{L}_m$ assumes small values. A simple system was chosen, and it is straightforward to extend the analysis to, say, three-dimensional geometries, such as a circular curved pipe. We have already mentioned the advantages of having close form solutions for the particle phase, especially when the motion of the fluid phase cannot be simplified as in the present case. Finally, we mention that the same analysis can be applied to systems where $\tilde{L}_m$ assumes a large value.

Acknowledgement

The author wishes to express his gratitude for the encouragement given him by A. V. LEVY, Program Advisor, and M. HOLT, Faculty Advisor. — This work was supported by the Division of Materials Sciences, Office of Basic Energy Sciences, U.S. Department of Energy under Contract No. W-7405-ENG-48.

References

1 YEUNG, W. S., Erosion in a Curved Pipe, LBL Report No. 7354 (1977), also see Wear, **55** (1979), 91.
2 YEUNG, W. S., Fundamentals of the particulate phase in a gas-solid mixture, LBL Report No 8440, Nov. 1978, Berkeley, Calif.
3 LANGEVIN, P., Comptes Rend. Acad. Sci. Paris **14b**, 530 (1908).
4 PATHRIA, R. K., Statistical Mechanics, Pergamon Press (1977).
5 VAN DYKE, Perturbation Methods in Fluid Mechanics, The Parabolic Press, Stanford, Calif. (1975).
6 ERDELYI, A., An expansion procedure for singular perturbations, Acad. Sci. Torino, Cl. Sci. Fis. Mat. Nat. **95** (1961).

Eingereicht am 2. 2. 1979

Anschrift: Dr. WOON-SHING-YEUNG, Materials and Molecular Research Division, Lawrence Berkeley Laboratory, University of California, Berkeley, California 94720, USA. Present address: Thayer School of Engineering, Dartmouth College, New Hampshire, 03755, USA.

ZAMM 60, 249–255 (1980)

H. Ramkissoon[1]) / S. R. Majumdar[2])

Potentials and Green's Functions in Micropolar Fluid Theory

In dieser Arbeit wird ein System von gekoppelten Differentialgleichungen untersucht, die die schleichende stationäre, inkompressible Strömung einer mikropolaren Flüssigkeit charakterisieren. Unter Verwendung der singulären Fundamentallösungen, die einer konzentrierten Kraft und einem konzentrierten Drehmoment entsprechen, wird eine Integraldarstellung für den Geschwindigkeitsvektor und den Mikrorotationsvektor gewonnen. Die besondere Gestalt dieser Integraldarstellungen legt Einführung von Oberflächenpotentialen der einfachen und der doppelten Schicht nahe. Die Eigenschaften dieser Potentiale werden durch mehrere Sätze dargelegt, jedoch ohne detaillerte Beweise, da sie den üblichen harmonischen Potentialen sehr ähnlich sind. Des weiteren werden dann unter Verwendung der singulären Fundamentallösungen Greensche Matrizenfunkticnen berechnet. Unter Verwendung der Integraldarstellungen werden die formalen Lösungen für zwei Arten von Randwertproblemen in der Gestalt von Integralen erhalten, welche die gegebenen Körperkräfte, Oberflächendaten und eben diese Greenschen Matrizenfunktionen enthalten.

The system of coupled differential equations characterizing the slow, steady, incompressible flow of micropolar fluid is examined. Integral representations for the velocity vector and the microrotation vector are obtained by utilizing the fundamental singular solutions corresponding to a concentrated force and a concentrated couple. The very form of these integral representations suggests the introduction of surface potentials of single and double layers. The properties of these potentials are stated in various theorems without detailed proofs since they behave much like the ordinary harmonic potentials. Matrix Green's functions are then constructed with the aid of the fundamental singular solutions. By making use of the integral representations, formal solutions to two types of boundary value problems are generated in terms of integrals which include given body forces, surface data and these matrix Green's functions.

В данной работе исследуется система связанных друг с другом дифференциальных уравнений, характеризующих медленный непрерывный и несжимаемый поток микрополярной жидкости. С помощью фундаментальных особых решений, соответствующих сосредоточенной силе и сосредоточенной паре, получено интегральное представление для вектора скорости и вектора микровращения. Особая форма этого интегрального представления позволяет введение поверхностных потенциалов простого и двойного слоя. Свойства этих потенциалов представлены в форме нескольких теорем без подробных доказательств, т.к. эти потенциалы сходны с обычными гармоническими потенциалами. Используя фундаментальные особые решения, расчитаны далее матричные функции Грина. С помощью интегральных представлений получены формальные решения в форме интегралов для двух видов задач с граничными условиями. Эти решения содержат заданные силы, данные о поверхности и вышеупомянутые матричные функции Грина.

1. Introduction

The consideration of couple stress, in addition to the classical Cauchy stress has led to the recent development of several theories [1], [2], [3], [4] of fluid microcontinua. This new branch of fluid mechanics, and in particular the specialized theory of micropolar fluid advanced by Eringen [1], has been under intensive investigation during recent years mainly because it possesses the mechanism to describe such rheologically complex fluids as liquid crystals, polymeric suspensions and animal blood for which the classical Navier-Stokes theory is inadequate.

The many and diverse applications of microcontinuum mechanics to flow problems are well documented in a review article by Ariman et al. [5]. Kline and Allen [6] studied blood flow based on a microcontinuum formulation by investigating the concentration effects in oscillatory blood flow. More recently, with an improved microcontinuum model [3] of blood, Ariman et al. [7] presented an encouraging comparison of the theoretical velocity and cell-rotational velocity profiles with the experimental data of Bugliarello and Sevilla [8].

As a result of these interesting findings, it is expected that further research work, both theoretical and experimental, will be undertaken in this new and challenging area. With this in mind, the following theoretical work is presented as foundation material for possible use in the study of existence and uniqueness of solutions.

This paper examines the system of equations governing the slow, steady, incompressible flow of micropolar fluids. The key point to note in the theory of these fluids, is the introduction of an independent kinematic vector, called the microrotation vector, to represent the spin velocity of the particles. Integral representations for both the velocity vector and the independent microrotation vector are obtained and their very form suggests the introduction of surface potentials. The properties of these potentials, due to the polarity of the Kernels, are analogous to the properties of potentials of single and double layers and they are therefore stated in the form of theorems without detailed proof. Finally, Green's functions are introduced, and the solutions to two fundamental boundary value problems are then generated.

2. Basic equations and integral representations

The equations of motion characterizing slow, steady flow of a micropolar fluid are given by [1]

$$t_{ji,j} + f_i = 0\,, \tag{1}$$

$$m_{ji,j} + \varepsilon_{ijk} t_{jk} + l_i = 0\,, \tag{2}$$

[1]) Department of Mathematics, University of the West Indies, St. Augustine, Trinidad, W. I.

[2]) Department of Mathematics, University of Calgary, Calgary, Alberta, Canada.

while the linear constitutive laws take the form

$$t_{ij} = t_{ij}(u, v) = (-p + \lambda u_{k,k})\,\delta_{ij} + \mu(u_{i,j} + u_{j,i}) + \varkappa(u_{j,i} - \varepsilon_{ijk}v_k)\,, \tag{3}$$

$$m_{ij} = m_{ij}(v) = \alpha v_{k,k}\delta_{ij} + \beta v_{i,j} + \gamma v_{j,i}\,. \tag{4}$$

In these equations

t_{ij} = components of the stress tensor,
m_{ij} = components of the couple stress tensor,
u_i = components of the velocity vector,
v_i = components of the microrotation vector,
f_i = components of the body force,
l_i = components of the body couple,
p = pressure
δ_{ij} = Kronecker delta,
ε_{ijk} = alternating tensor,

and $(\alpha, \beta, \gamma; \lambda, \mu, \varkappa)$ are constants characteristic of the particular fluid under consideration.

In the case of incompressible flows, the above equations (1)—(4) reduce to the following system of coupled differential equations. Continuity equation

$$\nabla \cdot \boldsymbol{u} = 0 \quad \text{in} \quad V\,; \tag{5}$$

Balance of momentum equation

$$-(\mu + \varkappa)\,\nabla \times \nabla \times \boldsymbol{u} + \varkappa\nabla \times \boldsymbol{v} - \nabla p + \boldsymbol{f} = 0 \quad \text{in} \quad V; \tag{6}$$

Balance of first stress moment equation

$$(\alpha + \beta + \gamma)\,\nabla\nabla \cdot \boldsymbol{v} - \gamma\nabla \times \nabla \times \boldsymbol{v} + \varkappa\nabla \times \boldsymbol{u} - 2\varkappa\boldsymbol{v} + \boldsymbol{l} = 0 \quad \text{in} \quad V. \tag{7}$$

It should be noted that the constant $\varkappa$ links together the velocity field $\boldsymbol{u}$ and the microrotation field $\boldsymbol{v}$ and is often termed the coupling constant. If this quantity $\varkappa$ vanishes, the classical Navier-Stokes theory is recovered for the velocity field.

For convenience, we define the body operators $\square^*$ and $\square$ as follows:

$$\square^*(\boldsymbol{u}, \boldsymbol{v}, p) = -(\mu + \varkappa)\,\nabla \times \nabla \times \boldsymbol{u} + \varkappa\nabla \times \boldsymbol{v} - \nabla p\,, \tag{8}$$

$$\square(\boldsymbol{u}, \boldsymbol{v}) = (\alpha + \beta + \gamma)\,\nabla\nabla \cdot \boldsymbol{v} - \gamma\nabla \times \nabla \times \boldsymbol{v} + \varkappa\nabla \times \boldsymbol{u} - 2\varkappa\boldsymbol{v}\,. \tag{9}$$

The following reciprocal theorem can be easily verified [9]:

Theorem 1. *Let* $(\boldsymbol{u}', \boldsymbol{v}', p', t'_{ij}, m'_{ij}, \boldsymbol{f}', \boldsymbol{l}')$ *and* $(\boldsymbol{u}'', \boldsymbol{v}'', p'', t''_{ij}, m''_{ij}, \boldsymbol{f}'', \boldsymbol{l}'')$ *represent any two motions of the same fluid which conform to equations* (1)—(4). *Let A be a closed surface bounding any fluid volume V and* $\boldsymbol{u}', \boldsymbol{v}', \boldsymbol{u}'', \boldsymbol{v}'' \in C^1$ *in* $A + V$. *It can then be shown that*

$$\int_A [t'_{ji}u''_i - t''_{ji}\,u'_i]n_j\,dA + \int_A [m'_{ji}v''_i - m''_{ji}v'_i]\,n_j\,dA = \tag{10}$$
$$= \int_V (\square^*(\boldsymbol{u}', \boldsymbol{v}', p') \cdot \boldsymbol{u}'' - \square^*(\boldsymbol{u}'', \boldsymbol{v}'', p'') \cdot \boldsymbol{u}']\,dV + \int_V [\square(\boldsymbol{u}', \boldsymbol{v}') \cdot \boldsymbol{v}'' - \square(\boldsymbol{u}'', \boldsymbol{v}'') \cdot \boldsymbol{v}']\,dV$$

where $\boldsymbol{n}$ *is the unit outward drawn normal vector.*

The proof of this theorem is given in a previous paper by the authors [9].

Let us now assume that the vector functions $\boldsymbol{u}, \boldsymbol{v}$ possess no singularity in the fluid region V and satisfy the non-homogeneous Stokes system

$$\square^*(\boldsymbol{u}, \boldsymbol{v}, p) = -\boldsymbol{f}(x)\,, \quad \square(\boldsymbol{u}, \boldsymbol{v}) = -\boldsymbol{l}(x)\,, \quad \nabla \cdot \boldsymbol{u} = 0 \qquad (11), (12), (13)$$

for $x \in V$. This is equivalent to the system of equations given by (5), (6) and (7). Let $({}_F\boldsymbol{u}^k, {}_F\boldsymbol{v}^k, {}_Fp^k)$ and $({}_L\boldsymbol{u}^k, {}_L\boldsymbol{v}^k, {}_Lp^k)$ be the fundamental singular solutions for a concentrated force and a concentrated couple respectively, applied at the point y. Hence

$$\square^*({}_F\boldsymbol{u}^k, {}_F\boldsymbol{v}^k, {}_Fp^k) = -\delta(x - y)\,\boldsymbol{e}^k, \quad \square({}_F\boldsymbol{u}^k, {}_F\boldsymbol{v}^k) = 0\,, \quad \nabla \cdot {}_F\boldsymbol{u}^k = 0\,, \qquad (14), (15), (16)$$

and

$$\square^*({}_L\boldsymbol{u}^k, {}_L\boldsymbol{v}^k, {}_Lp^k) = 0\,, \quad \square({}_L\boldsymbol{u}^k, {}_L\boldsymbol{v}^k) = -\delta(x - y)\,\boldsymbol{e}^k\,, \quad \nabla \cdot {}_L\boldsymbol{u}^k = 0\,, \qquad (17), (18), (19)$$

for $x \in E$ (infinite space) and $k = 1, 2, 3$. Here $\boldsymbol{e}^k$ is a unit vector defined along the kth co-ordinate axis, $\delta(x - y)$ is the Dirac delta function and all differentiations are carried out with respect to the variable x with y playing the role of a parameter. Applying the reciprocal theorem to the systems (11)—(13) and (14)—(16) and then (11)—(13) and (17)—(19) in turn, we obtain respectively the following integral representation theorems for $\boldsymbol{u}(x)$ and $\boldsymbol{v}(x)$:

Theorem 2. *Let ${}_Ft_{ij}{}^k$ and ${}_Fm_{ij}{}^k$ be the components of the stress tensor and the couple stress tensor, respectively, associated with the fundamental solutions due to the concentrated point force $\delta(x - y)\, e^k$. Let ${}_Lt_{ij}{}^k$ and ${}_Lm_{ij}{}^k$ be the corresponding quantities associated with the fundamental solutions due to the concentrated point couple $\delta(x - y)\, e^k$. The integral representations for $\boldsymbol{u}(x)$ and $\boldsymbol{v}(x)$ are given by*

$$\begin{aligned} u_k(x) &= \int_A [t_{ji}^{(y)}(\boldsymbol{u}, \boldsymbol{v})\, {}_Fu_i{}^k(x, y) - {}_Ft_{ji}{}^k(x, y)\, u_i(y)]\, n_j\, dA_y + \\ &+ \int_A [m_{ji}^{(y)}(\boldsymbol{v})\, {}_Fv_i{}^k(x, y) - {}_Fm_{ji}{}^k(x, y)\, v_i(y)]\, n_j\, dA_y \\ &+ \int_V [f_i(y)\, {}_Fu_i{}^k(x, y) + l_i(y)\, {}_Fv_i{}^k(x, y)]\, dV_y\,, \end{aligned} \tag{20}$$

$$\begin{aligned} v_k(x) &= \int_A [t_{ji}{}^{(y)}(\boldsymbol{u}, \boldsymbol{v})\, {}_Lu_i{}^k(x, y) - {}_Lt_{ji}{}^k(x, y)\, u_i(y)]\, n_j\, dA_y \\ &+ \int_A [m_{ji}{}^{(y)}(\boldsymbol{v})\, {}_Lv_i{}^k(x, y) - {}_Lm_{ji}{}^k(x, y)\, v_i(y)]\, n_j\, dA_y \\ &+ \int_V [f_i(y)\, {}_Lu_i{}^k(x, y) + l_i(y)\, {}_Lv_i{}^k(x, y)]\, dV_y\,. \end{aligned} \tag{21}$$

Corollary 1. *An integral representation for the pressure can be obtained if we substitute the expressions just derived into equation* (6).

Corollary 2. *In the absence of couple stresses and body couples, equation* (20) *reduces to the form*

$$u_k(x) = \int_A [t_{ji}{}^{(y)}(\boldsymbol{u})\, {}_Fu_i{}^k(x, y) - {}_Ft_{ji}{}^k(x, y)\, u_i(y)]\, n_j dA_y + \int_V f_i(y)\, {}_Fu_i{}^k(x, y)\, dV_y\,.$$

This is the classical result given by LADYZHENSKAYA [10].

Explicit expressions for the fundamental singular solutions have been obtained by the authors [9] and are given by:

$${}_F\boldsymbol{u}^k = \frac{1}{4\pi\sigma^2(\mu + \varkappa)} \left[\frac{\varkappa}{\gamma}\frac{\boldsymbol{e}^k}{r} + \frac{\varkappa}{\gamma}\frac{(\boldsymbol{e}^k \cdot \boldsymbol{r})}{r^3}\boldsymbol{r} + \frac{2\varkappa - \gamma\sigma^2}{\gamma\sigma^2}\left\{\nabla \times \nabla \times \boldsymbol{e}^k\, \frac{(1 - \mathrm{e}^{-\sigma r})}{r}\right\}\right], \tag{22}$$

$${}_F\boldsymbol{v}^k = \frac{1}{4\pi(2\mu + \varkappa)}\, \boldsymbol{e}^k \times \nabla\, \frac{(1 - \mathrm{e}^{-\sigma r})}{r}, \tag{23}$$

$${}_Fp^k = \frac{1}{4\pi}\frac{\boldsymbol{e}^k\ \boldsymbol{r}}{r^3}, \tag{24}$$

$${}_L\boldsymbol{u}^k = -\frac{1}{4\pi(2\mu + \varkappa)}\, \nabla \times \boldsymbol{e}^k\, \frac{(1 - \mathrm{e}^{-\sigma r})}{r}, \tag{25}$$

$$\begin{aligned} {}_L\boldsymbol{v}^k &= \frac{\boldsymbol{e}^k}{4\pi\gamma}\frac{\mathrm{e}^{-\sigma r}}{r}\, \frac{\nabla\nabla\ \boldsymbol{e}^k}{4\pi\gamma(\mu + \varkappa)\,(\alpha + \beta + \mu)} \times \\ &\times \left[\frac{\varkappa^2}{\sigma^2\tau^2}\frac{1}{r} + \frac{\varkappa^2 - \sigma^2\,(\alpha + \beta)\,(\mu + \varkappa)}{\sigma^2(\sigma^2 - \tau^2)}\frac{\mathrm{e}^{-\sigma r}}{r} + \frac{\varkappa^2 - \tau^2(\alpha + \beta)\,(\mu + \varkappa)}{\tau^2(\tau^2 - \lambda^2)}\frac{\mathrm{e}^{-\tau r}}{r}\right], \end{aligned} \tag{26}$$

$${}_Lp^k = 0\,, \tag{27}$$

where $\sigma^2 = \dfrac{\varkappa(2\mu + \varkappa)}{\gamma(\mu + \varkappa)}$, $\quad \tau^2 = \dfrac{2\varkappa}{\alpha + \beta + \gamma}$ $\quad$ and $\quad r^2 = \sum\limits_{i=1}^{3} (x_i - y_i)^2$.

Thus far, we gave generated integral representations for the velocity and microrotation vectors associated with the system of equations (11)—(13) which govern the slow, steady, incompressible flow of micropolar fluids. As in ordinary hydrodynamical potential theory [10], these representation formulae (20) and (21) suggest the introduction of surface potentials.

3. Potentials of single and double layers

Let

$$\boldsymbol{U}_0(x; \boldsymbol{\varphi}, \overline{\boldsymbol{\varphi}}) = \int_A [{}_F\boldsymbol{u}^k(x, y)\, \varphi_k(y) + {}_L\boldsymbol{u}^k(x, y)\, \overline{\varphi}_k(y)]\, dA_y\,, \tag{28}$$

$$\boldsymbol{W}_0(x; \boldsymbol{\varphi}, \overline{\boldsymbol{\varphi}}) = \int_A [{}_F\boldsymbol{v}^k(x, y)\, \varphi_k(y) + {}_L\boldsymbol{v}^k(x, y)\, \overline{\varphi}_k(y)]\, dA_y\,, \tag{29}$$

$$U_k(x; \boldsymbol{\psi}, \overline{\boldsymbol{\psi}}) = \int_A [{}_Ft_{ji}^k(x, y)\, \psi_i(\boldsymbol{y}) + {}_Lt_{ji}^k(x, y)\, \overline{\psi}_i(y)]\, n_j\, dA_y\,, \tag{30}$$

and

$$W_k(x; \boldsymbol{\psi}, \overline{\boldsymbol{\psi}}) = \int_A [{}_Fm_{ji}^k(x, y)\, \psi_i(y) + {}_Lm_{ji}^k(x, y)\, \overline{\psi}_i(y)]\, n_j\, dA_y\,. \tag{31}$$

The integrals given by (28) and (29) are single layer potentials while those given by (30) and (31) are double layer potentials. With the aid of these integrals, equations (20) and (21) can now be rewritten in the form

$$u_k(x) = U_{ok}(x, \boldsymbol{T}_j n_j, 0) - U_k(x; \boldsymbol{u}, 0) + W_{ok}(x; \boldsymbol{M}_j n_j, 0) - W_k(x; \boldsymbol{v}, 0) + \\ + \int_V [f_i(y)\, {}_F u_i^k(x, y) + l_i(y)\, {}_F v_i^k(x, y)]\, dV_y \tag{32}$$

and

$$v_k(x) = U_{ok}(x; 0, \boldsymbol{T}_j n_j) - U_k(x; 0, \boldsymbol{u}) + W_{ok}(x; 0, \boldsymbol{M}_j n_j) - W_k(x; 0, \boldsymbol{v})\,, \tag{33}$$

where

$$\boldsymbol{T}_j = [t_{1j}(\boldsymbol{u}, \boldsymbol{v}), t_{2j}(\boldsymbol{u}, \boldsymbol{v}), t_{3j}(\boldsymbol{u}, \boldsymbol{v})]\,,$$

and

$$\boldsymbol{M}_j = [m_{1j}(\boldsymbol{v}), m_{2j}(\boldsymbol{v}), m_{3j}(\boldsymbol{v})]\,.$$

We would now like to examine the behaviour of these potentials defined by equations (28)—(31) at the boundary or more precisely as the point at which they are evaluated approaches the surface A from inside and outside. To do so, we make the following assumptions:

(a) A is a closed LYAPUNOV surface [11] — that is, a closed surface with continuously turning tangent plane,

(b) The surface density functions are continuous or satisfy the HÖLDER condition with exponent $\lambda_1 \leqq 1$ [11] — that is, to belong to the class $H(\lambda_1)$,

(c) ξ, η are points on the surface A.

Then we have the following Theorem:

Theorem 3. *The single layer potentials $\boldsymbol{U}_0(x; \boldsymbol{\varphi}, \overline{\boldsymbol{\varphi}})$ and $\boldsymbol{W}_0(x; \boldsymbol{\varphi}, \overline{\boldsymbol{\varphi}})$ are continuous on the whole space E including the surface A for continuous vector density functions $\boldsymbol{\varphi}(y)$ and $\overline{\boldsymbol{\varphi}}(y)$.*

Proof. Both the functions $\boldsymbol{U}_0(x; \boldsymbol{\varphi}, \overline{\boldsymbol{\varphi}})$ and $\boldsymbol{W}_0(x; \boldsymbol{\varphi}, \overline{\boldsymbol{\varphi}})$ given respectively by (28) and (29) exist and are obviously continuous everywhere except possibly on A. To examine their behaviour on A, we must determine the polarity of their kernels. Expanding (22), (23), (25) and (26) in series form, we get

$${}_F u_i^k = \frac{\delta_{ik}}{8\pi(\mu + \varkappa)}\left[\frac{1}{r}\right] + \frac{2\mu + 3\varkappa}{8\pi(\mu + \varkappa)\,(2\mu + \varkappa)}\left[\frac{(x_i - y_i)\,(x_k - y_k)}{r^3}\right] + O[r^n]\,, \tag{34}$$

$${}_F v_i^k = O[r^n]\,, \quad {}_L u_i^k = O[r^n]\,, \tag{35), (36}$$

$${}_L v_i^k = \frac{(\alpha + \beta + 2\gamma)\,\delta_{ik}}{8\pi\gamma(\alpha + \beta + \gamma)}\left[\frac{1}{r}\right] + \frac{(\alpha + \beta)}{8\pi\gamma\,(\alpha + \beta + \gamma)}\left[\frac{(x_i - y_i)\,(x_k - y_k)}{r^3}\right] + O[r^n]\,, \tag{37}$$

where $n \geqq 0$. Hence the kernels of these potentials possess polarity $\frac{1}{r}$ and consequently they are continuous for points on A as well [12].

We shall now prove the following Theorem to display the discontinuity of the potential given by the integral (30).

Theorem 4. *Let $U_k(\xi; \boldsymbol{\psi}, \overline{\boldsymbol{\psi}})_{(i)}$ and $U_k(\xi; \boldsymbol{\psi}, \overline{\boldsymbol{\psi}})_{(e)}$ denote the limiting values of $U_k(x; \boldsymbol{\psi}, \overline{\boldsymbol{\psi}})$ on A as A is approched from inside and outside of V, respectively. If the vector density functions $\boldsymbol{\psi}$ and $\overline{\boldsymbol{\psi}} \in H(\lambda_1)$, then*

$$U_k(\xi; \boldsymbol{\psi}, \overline{\boldsymbol{\psi}})_{(i)} = \tfrac{1}{2}\psi_k(\xi) + \int_A [{}_F t_{ji}^k(\xi, \eta)\, \psi_i(\eta) + {}_L t_{ij}^k(\xi, \eta)\, \overline{\psi}_i(\eta)]\, n_j\, dA_\eta\,, \tag{38}$$

and

$$U_k(\xi; \boldsymbol{\psi}, \overline{\boldsymbol{\psi}})_{(e)} = -\tfrac{1}{2}\psi_k(\xi) + \int_A [{}_F t_{ji}^k(\xi, \eta)\, \psi_i(\eta) + {}_L t_{ij}^k(\xi, \eta)\, \overline{\psi}_i(\eta)]\, n_j\, dA_\eta\,. \tag{39}$$

Proof. These relations are analogous to those for jumps of the harmonic potential of a double layer. In order to display the discontinuity of the potential $U_k(x; \boldsymbol{\psi}, \overline{\boldsymbol{\psi}})$ on the surface A, we must, as in the previous theorem, seek power series expansions for the kernels. Using (3), (4), (22), (23), (25) and (26), we obtain the following after some working:

$${}_F t_{ij}^k = \frac{\varkappa\delta_{ij}}{8\pi(\mu + \varkappa)}\left[\frac{(x_k - y_k)}{r^3}\right] - \frac{\varkappa^2\delta_{jk}}{8\mu(\mu + \varkappa)\,(2\mu + \varkappa)}\left[\frac{(x_i - y_i)}{r^3}\right] + \\ + \frac{\varkappa(3\varkappa + 4\mu)\,\delta_{ki}}{8\pi(\mu + \varkappa)\,(2\mu + \varkappa)}\left[\frac{(x_j - y_j)}{r^3}\right] - \frac{3(2\mu + \varkappa)}{8\pi(\mu + \varkappa)}\left[\frac{(x_i - y_i)\,(x_j - y_j)\,(x_k - y_k)}{r^5}\right] + O[r^{n-1}]\,, \tag{40}$$

$${}_F m_{ij}^k = O[r^{n-1}]\,, \quad {}_L t_{ij}^k = O[r_{n-1}]\,, \tag{41), (42}$$

$${}_L m_{ij}^k = \frac{(\alpha + \beta)\,(\beta + \gamma)\,\delta_{ij}}{8\pi\gamma(\alpha + \beta + \gamma)}\left[\frac{(x_k - y_k)}{r^3}\right] + \frac{(\alpha\beta + \beta^2 - 2\gamma - \gamma\beta)\,\delta_{jk}}{8\pi\gamma(\alpha + \beta + \gamma)}\left[\frac{(x_i - y_i)}{r^3}\right] + \\ + \frac{(\alpha\gamma - \alpha\beta - \beta^2 - \gamma\beta)\,\delta_{ki}}{8\pi\gamma(\alpha + \beta + \gamma)}\left[\frac{(x_j - y_j)}{r^3}\right] - \frac{3(\alpha + \beta)\,(\gamma + \beta)}{8\pi\gamma(\alpha + \beta + \gamma)}\left[\frac{(x_i - y_i)(x_j - y_j)\,(x_k - y_k)}{r^5}\right] + O[r^{n-1}]. \tag{43}$$

where as before $n \geqq 0$. Thus ${}_Fm_{ij}^k$ and ${}_Lt_{ij}^k$ possess singularities of order $\frac{1}{r}$ while the singularities of ${}_Lm_{ij}^k$ and ${}_Ft_{ij}^k$ are of order $\frac{1}{r^2}$. In other words, the kernels of the potential $U_k(x;\boldsymbol{\psi},\overline{\boldsymbol{\psi}})$ have polarity of $\frac{1}{r^2}$ at most. Hence $U_k(x;\ \boldsymbol{\psi},\overline{\boldsymbol{\psi}})$ is not continuous in passing through A as it has the same kind of discontinuity on A as its counterpart in the classical NAVIER-STOKES theory. Consequently, the proof of the above results (38) and (39) is similar to that given by LADYZHENSKAYA [10] or any standard text on potential theory [12].

It is obvious that the potential $W_k(x;\boldsymbol{\psi},\overline{\boldsymbol{\psi}})$ has a similar jump at the passage through the surface A, i.e.

$$W_k(\xi;\boldsymbol{\psi},\overline{\boldsymbol{\psi}})_{(i)} = \tfrac{1}{2}\overline{\psi}_k(\xi) + \int_A [{}_Fm_{ji}^k(\xi;\eta)\,\psi_i(\eta) + {}_Lm_{ji}^k(\xi,\eta)\,\overline{\psi}_i(\eta)]\,n_j\,\mathrm{d}A_\eta\,, \tag{44}$$

$$W_k(\xi;\boldsymbol{\psi},\overline{\boldsymbol{\psi}},)_{(e)} = -\tfrac{1}{2}\overline{\psi}_k(\xi) + \int_A {}_Fm_{ji}^k(\xi,\eta)\,\psi_i(\eta) + {}_Lm_{ji}^k(\xi,\eta)\,\overline{\psi}_i(\eta)]\,n_j\,\mathrm{d}A_\eta. \tag{45}$$

Recall from Theorem 3 that $U_0(x;\boldsymbol{\varphi},\overline{\boldsymbol{\varphi}})$ and $W_0(x;\boldsymbol{\varphi},\overline{\boldsymbol{\varphi}})$ are both continuous throughout E. Consider now the corresponding stress tensor $t_{ij}(\boldsymbol{U}_0, \boldsymbol{w}_0)$ and couple stress tensor $m_{ij}(\boldsymbol{W}_0)$ defined in (3) and (4), respectively. As x approaches the point ξ along the normal $\boldsymbol{n}$ to A either from the outside or the inside of A, the functions $t_{ij}(\boldsymbol{U}_0,\boldsymbol{W}_0)\,n_j$ and $m_{ij}(\boldsymbol{W}_0)n_j$ have well defined limiting values. Moreover, following the procedure used in proving Theorem 4, we can likewise establish the following results:

Theorem 5. *For vector density functions $\boldsymbol{\varphi},\overline{\boldsymbol{\varphi}} \in H(\lambda_1)$, the functions $t_{ij}(\boldsymbol{U}_0,\ \boldsymbol{W}_0)n_j$ and $m_{ij}(\boldsymbol{W}_0)n_j$ tend to the following finite limits as x approaches the boundary point $\xi \in A$ from the inside and from the outside:*

$$t_{ij}(\boldsymbol{U}_0,\boldsymbol{W}_0)_{(i)}\,n_j = \tfrac{1}{2}\varphi_i(\xi) + \int_A t_{ij}\,[{}_F\boldsymbol{u}^k(\xi,\eta)\,\varphi_k(\eta) + {}_L\boldsymbol{u}^k(\xi,\eta)\,\overline{\varphi}_k(\eta) + {}_F\boldsymbol{v}^k(\xi,\eta)\,\varphi_k(\eta) + {}_L\boldsymbol{v}^k(\xi,\eta)\,\overline{\varphi}_k(\eta)]\,n_j\,\mathrm{d}A_\eta\,, \tag{46}$$

$$t_{ij}(\boldsymbol{U}_0,\boldsymbol{W}_0)_{(e)}\,n_j = -\tfrac{1}{2}\varphi_i(\xi) + \int_A t_{ij}[{}_F\boldsymbol{u}^k(\xi,\eta)\,\varphi_k(\eta) + {}_L\boldsymbol{u}^k(\xi,\eta)\,\overline{\varphi}_k(\eta) + {}_F\boldsymbol{v}^k(\xi,\eta)\,\varphi_k(\eta) + {}_L\boldsymbol{v}^k(\xi,\eta)\,\overline{\varphi}_k(\eta)]\,n_j\,\mathrm{d}A_\eta\,, \tag{47}$$

$$m_{ij}(\boldsymbol{W}_0)_{(i)}\,n_j = \tfrac{1}{2}\,\overline{\varphi}_i(\xi) + \int_A m_{ij}\,[{}_F\boldsymbol{v}^k(\xi,\eta)\,\varphi_k(\eta) + {}_L\boldsymbol{v}^k(\xi,\eta)\,\overline{\varphi}_k(\eta)\,]\,n_j\,\mathrm{d}A_\eta\,, \tag{48}$$

$$m_{ij}(\boldsymbol{W}_0)_{(e)}\,n_j = -\tfrac{1}{2}\,\overline{\varphi}_i(\xi) + \int_A m_{ij}\,[{}_F\boldsymbol{v}^k(\xi,\eta)\,\varphi_k(\eta) + {}_L\boldsymbol{v}^k(\xi,\eta)\,\overline{\varphi}_k(\eta)]\,n_j\,\mathrm{d}A_\eta\,. \tag{49}$$

The properties of the potentials which we have just established, should enable us to examine the DIRICHLET and NEUMANN type problems for any domains which are bounded by sufficiently smooth surfaces by reducing the problems to the form of integral equations as has been the case in both classical hydrodynamics [10] and elasticity [11].

4. The Green's Functions

With the aid of the fundamental singular solutions given by (22)–(27), the matrix GREEN's functions will now be constructed. These functions will then be utilized in the integral representations (20) and (21) to yield formulae for the velocity and microrotation vectors for certain boundary value problems in terms of the surface and volume integrals depending upon surface and volume data and the matrix GREEN's functions. Let us define the following matrix GREEN's functions:

$${}_FG_{ij}(x,y) = {}_Fu_i^j(x,y) - {}_Fg_i^j(x,y)\,, \qquad {}_FH_{ij}(x,y) = {}_Fv_i^j(x,y) - {}_Fh_i^j(x,y)\,, \tag{50), (51}$$

with

$${}_FQ_j(x,y) = {}_Fp^j(x,y) - {}_Fq^j(x,y)\,, \tag{52}$$

and

$${}_LG_{ij}(x,y) = {}_Lu_i^j(x,y) - {}_Lg_i^j(x,y)\,, \quad {}_LH_{ij}(x,y) = {}_Lv_i^j(x,y) - {}_Lh_i^j(x,y)\,, \tag{53), (54}$$

with

$${}_LQ_j(x,y) = {}_Lp^j(x,y) - {}_Lq^j(x,y)\,, \tag{55}$$

where ${}_Mu_i^j$, ${}_Mv_i^j$, ${}_Mp^j (M = F, L)$ are the fundamental solutions given by (22) – (27), and ${}_Mg_i^j$, ${}_Mh_i^j$, ${}_Mq^j$ are defined as the solutions of the STOKES problem:

$$\square^*({}_M\boldsymbol{g}^j,\ {}_M\boldsymbol{h}^j,\ {}_Mq^j) = 0,\quad \square({}_M\boldsymbol{g}^j,\ {}_M\boldsymbol{h}^j) = 0,\quad \nabla\cdot{}_M\boldsymbol{g}^j = 0\,, \tag{56), (57), (58}$$

and satisfy certain boundary conditions specified for each boundary value problem.

On applying the reciprocal theorem to the system $(\boldsymbol{u},\boldsymbol{v},p)$ and $({}_M\boldsymbol{g}^j,\ {}_M\boldsymbol{h}^j,\ {}_Mq^j)$ associated respectively with (11)–(13) and (56)–(58), we obtain:

$$\int_A [t_{ji}^{(y)}(\boldsymbol{u},\boldsymbol{v})\ {}_Mg_i^k(x,y) - {}_M\overline{t}_{ji}^k(x,y)\ u_i(y)]\,n_j\,\mathrm{d}A_y + \int_A [m_{ji}^{(y)}(\boldsymbol{v})\ {}_Mh_i^k(x,y) - {}_M\overline{m}_{ji}{}^k(x,y)\,v_i(y)]\,n_j\,\mathrm{d}A_y + \int_V [f_i(y)\ {}_Mg_i^k(x,y) + l_i(y)\ {}_Mh_i^k(x,y)]\,\mathrm{d}V_y = 0\,, \tag{59}$$

where ${}_M\bar{t}_{ji}^k$ and ${}_M\bar{m}_{ji}^k$ are associated with the system (56) — (58). With the aid of (20), (21), (50) — (55) and (59), we obtain the following integral representations for the velocity and microrotation vectors:

$$u_k(x) = \int_A [t_{ji}^{(y)}(\boldsymbol{u}, \boldsymbol{\nu})\ {}_FG_{ik}(x, y) - {}_Ft_{ji}^{*k}(x, y)\ u_i(y)]\ n_j\, \mathrm{d}A_y + \tag{60}$$

$$+ \int_A [m_{ji}^{(y)}(\boldsymbol{\nu})\ {}_FH_{ik}(x, y) - {}_Fm_{ji}^{*k}(x, y)\ \nu_i(y)]\ n_j\, \mathrm{d}A_y + \int_V [f_i(y)\ {}_FG_{ik}(x, y) + l_i(y)\ {}_FH_{ik}(x, y)]\, \mathrm{d}V_y\,,$$

$$\nu_k(x) = \int_A [t_{ji}^{(y)}(\boldsymbol{u}, \boldsymbol{\nu})\ {}_LG_{ik}(x, y) - {}_Lt_{ji}^{*k}(x, y)\ u_i(y)]\ n_j\, \mathrm{d}A_y +$$

$$+ \int_A [m_{ji}^{(y)}(\boldsymbol{\nu})\ {}_LH_{ik}(x, y) - {}_Lm_{ji}^{*k}(x, y)\ \nu_i(y)]\ n_j\, \mathrm{d}A_y + \int_V [f_i(y)\ {}_LG_{ik}(x, y) + l_i(y)\ {}_LH_{ik}(x, y)]\, \mathrm{d}V_y\,, \tag{61}$$

where

$${}_Mt_{ji}^{*k} = {}_Mt_{ji}^k - {}_M\bar{t}_{ji}^k\,, \qquad {}_Mm_{ji}^{*k} = {}_Mm_{ji}^k - {}_M\bar{m}_{ji}^k\,.$$

These then are the general integral solutions of the non-homogeneous STOKES system (11) — (13), that is under investigation, in terms of the matrix GREEN's functions. It should be noted that these matrix GREEN's functions can be evaluated in approximately the same way as the fundamental singular solutions and as a result ${}_MG_{ij}$, ${}_MH_{ij}$, as well as ${}_Mt_{ij}^{*k}$ and ${}_Mm_{ij}^{*k}$, can be assumed to be known quantities. Let us now examine two boundary value problems.

(a) DIRICHLET *type problem.* If the vectors $\boldsymbol{u}, \boldsymbol{\nu}$ are prescribed on the surface A, and if ${}_Mg_i{}^j$, ${}_Mh_i{}^j$ satisfy the surface conditions

$${}_Mg_i{}^j(x, y)\,|_{x\in A} = {}_Ku_i{}^j(x, y)\,|_{x\in A}\,, \quad {}_Mh_i{}^j(x, y)\,|_{x\in A} = {}_M\nu_i{}^j(x, y)\,|_{x\in A}\,, \tag{62}$$

then from (60) and (61), the solutions of this DIRICHLET type boundary value problem is given by:

$$u_k(x) = -\int_A [{}_Ft_{ji}^{*k}(x, y)\ u_i(y) + {}_Fm_{ji}^{*k}(x, y)\ \nu_i(y)] n_j\, \mathrm{d}A_y + \int_V [f_i(y)\ {}_FG_{ik}(x, y) + l_i(y)\ {}_FH_{ik}(x, y)]\, \mathrm{d}V_y\,, \tag{63}$$

$$\nu_k(x) = -\int_A [{}_Lt_{ji}^{*k}(x, y)\ u_i(y) + {}_Lm_{ji}^{*k}(x, y)\ \nu_i(y)] n_j\, \mathrm{d}A_y + \int_V [f_i(y)\, {}_LG_{ik}(x, y) + l_i(y)\ {}_LH_{ik}(x, y)]\, \mathrm{d}V_y\,. \tag{64}$$

In the case of STOKES flow past an obstacle, if we make the assumption of adherence of micropolar fluid to the solid boundary, i.e.

$$\boldsymbol{u} = 0\,, \quad \boldsymbol{\nu} = 0 \quad \text{on} \quad A\,,$$

then (63) and (64) reduce respectively to:

$$u_k(x) = \int_V [f_i(y)\ {}_FG_{ik}(x, y) + l_i(y)\ {}_FH_{ik}(x, y)]\, \mathrm{d}V_y\,, \tag{65}$$

$$\nu_k(x) = \int_V [f_i(y)\ {}_LG_{ik}(x, y) + l_i(y)\ {}_LH_{ik}(x, y)]\, \mathrm{d}V_y\,. \tag{66}$$

Hence (65) and (66) represent the solutions of the STOKES problem (11)—(13) with zero boundary values of $\boldsymbol{u}$ and $\boldsymbol{\nu}$. In the absence of body couples, (65) becomes:

$$u_k(x) = \int_V f_i(y)\, {}_FG_{ik}(x, y)\, \mathrm{d}V_y\,. \tag{67}$$

This is the well-known formula for the classical case [10].

(b) NEUMANN *type problem.* In this type of boundary value problem, the boundary tractions and moments are prescribed on the surface, i.e.

$$t_{ji}^{(y)}(\boldsymbol{u}, \boldsymbol{\nu})\ n_j = t_i(y)\,, \quad m_{ji}^{(y)}(\boldsymbol{\nu})\ n_j = m_i(y)\,. \tag{68}$$

If we now construct the matrix GREEN's functions in such a way that

$${}_Mt_{ji}^k(x, y) = {}_M\bar{t}_{ji}^k(x, y), \quad {}_M\bar{m}_{ji}^k(x, y) = {}_Mm_{ji}^k(x, y)\,, \quad x \in A\,, \tag{69}$$

thus utilizing (60) and (61), the integral representations are:

$$u_k(x) = \int_A [t_i(y)\ {}_FG_{ik}(x, y) + m_i(y)\ {}_FH_{ik}(x, y)]\, \mathrm{d}A_y + \int_V [f_i(y)\ {}_FG_{ik}(x, y) + l_i(y)\ {}_FH_{ik}(x, y)]\, \mathrm{d}V_y\,, \tag{70}$$

$$\nu_k(x) = \int_A [t_i(y)\ {}_FG_{ik}(x, y) + m_i(y)\ {}_LH_{ik}(x, y)]\, \mathrm{d}A_y + \int_V [f_i(y)\ {}_LG_{ik}(x, y) + l_i(y)\ {}_LH_{ik}(x, y)]\, \mathrm{d}V_y\,. \tag{71}$$

To recover the corresponding result for the NAVIER-STOKES theory, we put $m_i \equiv 0$, $l_i = 0$ in (70), thus getting the velocity field as

$$u_k(x) = \int_A t_i(y)\ {}_FG_{ik}(x, y)\, \mathrm{d}A_y + \int_V f_i(y)\ {}_FG_{ik}(x, y)\, \mathrm{d}V_y\,. \tag{72}$$

5. Discussion

The main objective of this work has been to acquire basic theoretical results that may assist us in answering the fundamental question of uniqueness of solution of STOKES problem for a micropolar fluid together with suitable prescribed boundary conditions.

Integral representations for both the velocity and microrotation vectors have been constructed with the aid of fundamental singular solutions and a reciprocal theorem. These representations lead to the introduction of surface potentials which because of the polarity of their kernels, behave like ordinary potentials. In the case of

Newtonian fluids, similar results have been obtained by LADYZHENSKAYA [10] and have been utilized in discussing the question of uniqueness of the solution to the so-called STOKES problem. Finally, matrix GREEN's functions are introduced and formal solutions to two fundamental boundary value problems are expressed by integral representations involving these matrix GREEN's functions.

Acknowledgement

This research project was supported by the National Research Council of Canada (Grant No. A5294).

References

1 ERINGEN, A. C., J. Maths. Mech. **16**, 1 (1966).
2 STOKES, V. K., Phys. Fluids. 9, 1709 (1966).
3 AERO, E. L., BULYGIN, A. N., KUVSHINSKII, E. V., Prikl. Mat. Mekh. 29, 297 (1965), J. Appl. Math. Mech. 29, 333 (1965).
4 BLEUSTEIN, J. L., GREEN, A. E., Int. J. Engng. Sci. 5, 323 (1967).
5 ARIMAN, T., TURK, M. A., SYLVESTER, N. D., Int. J. Engng. Sci. 12, 273 (1974).
6 KLINE, K. A., ALLEN, S. J., Biorheology 6, 99 (1969).
7 ARIMAN, T., TURK, M. A., SYLVESTER, N. D., J. Appl. Mech. **41**, 1 (1974).
8 BUGLIARELLO, G., SEVILLA, J., Biorheology 7, 85 (1970).
9 RAMKISSOON, H., MAJUMDAR, S. R., Zeit. für Ang. Math. und Mech. (ZAMM) 56, 197 (1976).
10 LADYZHENSKAYA, O. A., The Mathematical Theory of Viscous Incompressible Flow, Gordon and Breach, Sci. Pub. Ltd., New York/London, 1969.
11 KUPRADZE, V. D., Israel Program for Scientific Translations, Jerusalem (1965).
12 SOBOLEV, S. L., Partial Differential Equations of Mathematical Physics, Addison-Wesley Pub. Co. Inc., London 1964.

Eingereicht am 19. 9. 1978

Anschrift: Dr. S. R. MAJUMDAR, The University of Calgary, Faculty of Science, 2920 24 AVE.N.W., Calgary, Canada, T2N 1N4

ZAMM 60, 257 –261 (1980)

H. Friedrich / K. Hennig

Numerische Auswertung von Berechnungsformeln für Überschreitenswahrscheinlichkeiten stochastisch beanspruchter mechanischer Systeme

Bei der Behandlung von stochastisch beanspruchten Konstruktionen hat die Berechnung der Überschreitenswahrscheinlichkeit — das ist die Wahrscheinlichkeit dafür, daß ein betrachteter stochastischer Prozeß in einem bestimmten Zeitintervall ein vorgegebenes Sicherheitsniveau mindestens einmal überschreitet — große praktische Bedeutung. Es wird auf die numerische Auswertung früher hergeleiteter Berechnungsformeln und einige Anwendungsbeispiele eingegangen.

For the treatment of the reliability and safety problems of random loaded constructions it is important to determine the first crossing probability. This means the probability that the construction failes if the considered stochastic process crosses a fixed level in a determined time interval at least one time. The numerical analysis of earlier derived computation formulas is considered.

При изучении конструкций со случайной нагрузкой большое практическое значение имеет рассчёт вероятности пересечения — это вероятность того, что рассматриваемый случайный процесс в определённом интервале времени пересечёт по крайней мере один раз заданный уровень надёжности. В работе обсуждаются численные рассчёты ранее полученных формул и некоторые примеры приложений.

1. Einleitung

Bei der Behandlung von stochastisch beanspruchten mechanischen Systemen hat die Berechnung der Überschreitenswahrscheinlichkeit $P_1(Z_0; 0, T)$ — das ist die Wahrscheinlichkeit dafür, daß ein betrachteter stochastischer Prozeß $Z(t)$ im Zeitintervall $[0, T)$ ein vorgegebenes konstantes Sicherheitsniveau Z_0 mindestens einmal überschreitet — große praktische Bedeutung, was sich in einer Fülle von Arbeiten auf diesem Gebiet zeigt (s. u. a. Heinrich-Hennig [1], Lin [2], Yang-Shinozuka [3]). Nachdem in einer vorangegangenen Arbeit (s. Friedrich-Hennig [4]) analytische Berechnungsformeln für die Überschreitenswahrscheinlichkeit hergeleitet worden sind, wird im folgenden auf deren numerische Auswertung eingegangen.

Von dem betrachteten stochastischen Prozeß $Z(t)$ setzen wir voraus, daß mit Wahrscheinlichkeit 1 die Ableitung $\dot{Z}(t)$ des Prozesses existiert. Die stochastischen Prozesse $Z(t)$, $\dot{Z}(t)$ und ihre gegenseitigen Abhängigkeiten seien durch die Folge ihrer Wahrscheinlichkeitsdichtefunktionen

$$f_{Z\dots Z\dot{Z}\dots\dot{Z}}(z_1, \dots, z_n, \dot{z}_1, \dots, \dot{z}_m; t_1, \dots, t_n, \bar{t}_1, \dots, \bar{t}_m), \qquad n, m = 0, 1, 2, \dots \tag{1}$$

bestimmt. Einige der in [4] gewonnenen Berechnungsformeln für die Überschreitenswahrscheinlichkeit $P_1(Z_0; 0, T)$, die im folgenden weiterbenutzt werden, sollen kurz angeführt werden:

$$P_1(Z_0; 0, T) = \sum_{k=1}^{\infty} (-1)^{k+1} \int\limits_{0 \leq \tau_1 < \dots < \tau_k \leq T} (k) \int p_k(\tau_1, \dots, \tau_k)\, d\tau_1 \dots d\tau_k = \sum_{k=1}^{\infty} \frac{(-1)^{k+1}}{k!} \int\limits_0^T (k) \int q_k(\tau_1, \dots, \tau_k)\, d\tau_1 \dots d\tau_k \tag{2}$$

mit

$$q_k(\tau_1, \dots, \tau_k) = \begin{cases} p_k(\tau_1, \dots, \tau_k) & \text{für nichtzusammenfallende Argumente} \\ \prod_{i=1}^{k} p_1(\tau_i) & \text{für mindestens zwei zusammenfallende Argumente} \end{cases}$$

und

$$p_k(\tau_1, \dots, \tau_k) = \int\limits_0^{\infty} (k) \int \dot{z}_1 \dots \dot{z}_k f_{Z\dots Z\dot{Z}\dots\dot{Z}}(Z_0, \dots, Z_0, \dot{z}_1, \dots, \dot{z}_k; \tau_1, \dots, \tau_k, \tau_1, \dots, \tau_k)\, d\dot{z}_1 \dots d\dot{z}_k \tag{3}$$

für nichtzusammenfallende Argumente $\tau_1, \dots, \tau_k$.

2. Numerische Bestimmung von Häufigkeiten

In praktischen Problemen werden der zu untersuchende stochastische Prozeß $Z(t)$ und dessen zeitliche Ableitung $\dot{Z}(t)$ durch Messungen oder numerische Berechnungen erhalten. Die entsprechenden Meßwerte zu den äquidistanten Zeitpunkten $t_i = (i - 1) \cdot \Delta t$ seien mit $z(t_i)$ und $\dot{z}(t_i)$, $i = 1, \dots, M$ bezeichnet. Falls die $\dot{z}(t_i)$-Werte nicht gemessen werden können, lassen sie sich näherungsweise durch numerische Differentiationsverfahren ermitteln. Um Formel (3) zur Berechnung von $p_k(\tau_1, \dots, \tau_k)$ anwenden zu können, ist es erforderlich, aus den vorliegenden Meßwerten $z(t_i)$ und $\dot{z}(t_i)$, $i = 1, \dots, M$ des stochastischen Prozesses $Z(t)$ und seiner Ableitung $\dot{Z}(t)$ zu diskreten Zeitpunkten t_i die Verteilungsdichtefunktionen

$$f_{Z\dots Z\dot{Z}\dots\dot{Z}}(Z_0, \dots, Z_0, \dot{z}_1, \dots, \dot{z}_k; t_1, \dots, t_k, t_1, \dots, t_k) \tag{4}$$

zu ermitteln. Für einen ergodischen stochastischen Prozeß $Z(t)$ wird die numerische Bestimmung von $f_{Z\dot{Z}}(Z_0, \bar{z}_1; 0, 0)$ und $f_{ZZ\dot{Z}\dot{Z}}(Z_0, Z_0, \bar{z}_1, \bar{z}_2; 0, \tau, 0, \tau)$ und von p_1 bzw. $p_2(\tau)$ angeführt. Damit lassen sich von (2) ausgehend Näherungen 1. und 2. Ordnung (s. FRIEDRICH-HENNIG [4]) bestimmen. Das beschriebene Vorgehen kann ohne größere Schwierigkeiten auf allgemeinere Fälle übertragen werden.

Es sei vorausgesetzt, daß der ergodische stochastische Prozeß $Z(t)$ und seine Ableitung $\dot{Z}(t)$ stetige Verteilungsdichtefunktionen $f_{Z\dot{Z}}(z_1, \bar{z}_1; 0, 0)$ und $f_{ZZ\dot{Z}\dot{Z}}(z_1, z_2, \bar{z}_1, \bar{z}_2; 0, \tau, 0, \tau)$ besitzen. Dann kann für $\Delta_1 \to 0$, $\Delta_2 \to 0$, $\bar{\Delta}_1 \to 0$, $\bar{\Delta}_2 \to 0$ näherungsweise gesetzt werden

$$f_{Z\dot{Z}}(z_1, \bar{z}_1; 0, 0)\, \Delta_1\bar{\Delta}_1 \approx P\{z_1 \leqq Z(t) < z_1 + \Delta_1, \bar{z}_1 \leqq \dot{Z}(t) < \bar{z}_1 + \bar{\Delta}_1\}$$

bzw. (5)

$$f_{ZZ\dot{Z}\dot{Z}}(z_1, z_2, \bar{z}_1, \bar{z}_2; 0, \tau, 0, \tau)\, \Delta_1\Delta_2\bar{\Delta}_1\bar{\Delta}_2 \approx$$

$$\approx P\{z_1 \leqq Z(t) < z_1 + \Delta_1, z_2 \leqq Z(t+\tau) < z_2 + \Delta_2, \bar{z}_1 \leqq \dot{Z}(t) < \bar{z}_1 + \bar{\Delta}_1, \bar{z}_2 \leqq \dot{Z}(t+\tau) < \bar{z}_2 + \bar{\Delta}_2\}\,.$$

Wir ersetzen weiterhin näherungsweise die Wahrscheinlichkeit P durch entsprechende relative Häufigkeiten. Bei den in (3) benötigten Verteilungsdichtefunktionen ist $z_1 = Z_0$ bzw. $z_1 = Z_0$ und $z_2 = Z_0$. Außerdem werden sie nur für $\bar{z}_1 \geqq 0$ bzw. $\bar{z}_1 \geqq 0$ und $\bar{z}_2 \geqq 0$ benötigt.

Wir zerlegen die positive $\bar{z}_1$-Achse in hinreichend viele Intervalle $[\bar{z}_{1,k-1}, \bar{z}_{1,k})$, $k = 1, \ldots, L_1$, die sich nicht überlappen, aber den endlichen Abschnitt $[0, \bar{z}_{1,L_1})$ ausfüllen, wie in Bild 1 skizziert ist. Die Intervallängen $\bar{z}_{1,k} - \bar{z}_{1,k-1}$, $k = 1, \ldots, L_1$ können variabel sein. Bei vorausgesetzter Ergodizität der stochastischen Prozesse $Z(t)$ und $\dot{Z}(t)$ liege eine Realisierung $z(t_i)$ von $Z(t)$ und eine entsprechende Realisierung $\dot{z}(t_i)$ von $\dot{Z}(t)$ in Form von M gemessenen bzw. berechneten Werten an den diskreten Zeitpunkten $t_i = (i = 1) \cdot \Delta t$, $i = 1, \ldots, M$ vor. Es bezeichne

$$\dot{z}^*(t_i) = \dot{z}(t_{i-1}) + \frac{\dot{z}(t_i) - \dot{z}(t_{i-1})}{z(t_i) - z(t_{i-1})}\{Z_0 - z(t_{i-1})\}$$

und $H^M_{Z_0}(k)$ die absolute Häufigkeit der Meßpunkte t_i mit $z(t_i) > Z_0$, $z(t_{i-1}) \leqq Z_0$ und $\bar{z}_{1,k-1} \leqq \dot{z}^*(t_i) < \bar{z}_{1,k}$ für $i = 1, \ldots, M$. Dann gilt bei hinreichend großem M

$$f_{Z\dot{Z}}(Z_0, \bar{z}_1; 0, 0) \approx \frac{H^M_{Z_0}(k)}{M \cdot (\bar{z}_{1,k} - \bar{z}_{1,k-1})} \quad \text{für} \quad \bar{z}_{1,k-1} \leqq \bar{z}_1 < \bar{z}_{1,k}\,, \qquad k = 1, \ldots, L_1\,. \tag{6}$$

Auf ähnliche Weise werden die positive $\bar{z}_1$-Achse in die Intervalle $[\bar{z}_{1,k-1}, \bar{z}_{1,k})$, $k = 1, \ldots, L_1$ und die positive $\bar{z}_2$-Achse in die Intervalle $[\bar{z}_{2,l-1}, \bar{z}_{2,l})$, $l = 1, \ldots, L_2$ unterteilt. Mit $\tau = s \cdot \Delta\tau$, $\Delta\tau = n \cdot \Delta t$, $s = 1, 2, \ldots$ bezeichne $H^M_{Z_0}(k, l)$ die absolute Häufigkeit der Meßpunkte t_i mit $z(t_i) > Z_0$, $z(t_{i-1}) \leqq Z_0$ und $z(t_i + \tau) > Z_0$, $z(t_{i-1} + \tau) \leqq Z_0$, sowie mit $\bar{z}_{1,k-1} \leqq \dot{z}^*(t_i) < \bar{z}_{1,k}$ und $\bar{z}_{2,l-1} \leqq \dot{z}^*(t_i + \tau) < \bar{z}_{2,l}$ für $i = 1, \ldots, M$. Dann folgt für hinreichend großes M

$$f_{ZZ\dot{Z}\dot{Z}}(Z_0, Z_0, \bar{z}_1, \bar{z}_2; 0, \tau, 0, \tau) \approx \frac{H^M_{Z_0}(k, l)}{M \cdot (\bar{z}_{1,k} - \bar{z}_{1,k-1})\,(\bar{z}_{2,l} - \bar{z}_{2,l-1})} \tag{7}$$

$$\text{für} \quad \bar{z}_{1,k-1} \leqq \bar{z}_1 < \bar{z}_{1,k};\ \bar{z}_{2,l-1} \leqq \bar{z}_2 < \bar{z}_{2,l}\,, \qquad k = 1, \ldots, L_1\,, \qquad l = 1, \ldots, L_2\,.$$

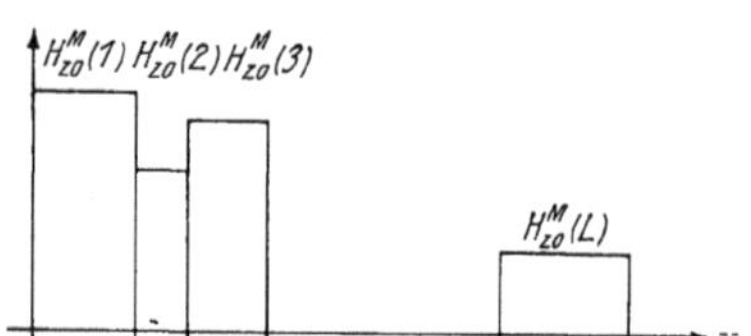

Bild 1. Absolute Häufigkeiten $H^M_{Z_0}(k)$ der Meßpunkte t_i mit $z(t_i) > Z_0$, $z(t_{i-1}) \leqq Z_0$ und $\bar{z}_{1,k-1} \leqq \dot{z}^*(t_i) < \bar{z}_{1,k}$

3. Auswertung unendlicher Integrale zur Berechnung von Verteilungsfunktionen

Die Auswertung der unendlichen Integrale in (3) zur Bestimmung von p_1 bzw. $p_2(\tau)$ erfolgt nach einem von SALZER-ZUCKER [5] vorgeschlagenen Vorgehen unter Benutzung der Laguerreschen Polynome. Dabei wird gesetzt

$$\int_0^\infty F(y)\, dy \approx \sum_{i=1}^{N} g_i^N F(y_i^N)\,; \tag{8}$$

y_i^N sind die Nullstellen des N-ten Laguerreschen Polynoms $L_N(y)$ und g_i^N Gewichtsfunktionen, die ebenso wie die y_i^N in Tabellenform vorliegen. Bei der Unterteilung der positiven $\bar{z}_1$-Achse achten wir darauf, daß die Punkte $\bar{z}_{1,k}$ $(k = 1, \ldots, L)$ mit den Nullstellen y_k^L des Laguerreschen Polynoms L-ten Grades zusammenfallen. Damit ergibt sich

$$p_1 = \int_0^\infty \bar{z}_1 f_{Z\dot{Z}}(Z_0, \bar{z}_1; 0, 0)\, d\bar{z}_1 = \frac{1}{M} \sum_{k=1}^{L_1} \frac{g_k^{L_1} y_k^{L_1} H^M_{Z_0}(k)}{(y_k^{L_1} - y_{k-1}^{L_1})}\,. \tag{9}$$

Setzt man bei der Unterteilung der positiven $\bar{z}_1$- und $\bar{z}_2$-Achse die Punkte $\bar{z}_{1,k}$ und $\bar{z}_{2,l}$ gleichfalls gleich den Nullstellen $y_k^{L_1}$ und $y_l^{L_2}$ der gewählten Laguerreschen Polynome vom Grade L_1 bzw. L_2, und läßt man für τ die diskreten

Werte $\tau = s \cdot \Delta\tau$, $\Delta\tau = n \cdot \Delta t$, $s = 1, 2, \dots$ zu, wenn $t_i = (i-1) \cdot \Delta t$, $i = 1, \dots, M$ ist, so folgt

$$p_2(\tau) = \int_0^\infty\int \bar{z}_1\bar{z}_2 f z z \dot{z}\dot{z}(Z_0, Z_0, \bar{z}_1, \bar{z}_2; 0, \tau, 0, \tau)\, d\bar{z}_1\, d\bar{z}_2 = \frac{1}{M}\sum_{k=1}^{L_1}\sum_{l=1}^{L_2} \frac{g_k^{L_1} g_l^{L_2} y_k^{L_1} y_l^{L_2} H_{Z_0}^M(k,l)}{(y_k^{L_1} - y_{k-1}^{L_1})(y_l^{L_2} - y_{l-1}^{L_2})}\,. \tag{10}$$

4. Numerische Berechnung von Überschreitenswahrscheinlichkeiten

Zur näherungsweisen Berechnung der Überschreitenswahrscheinlichkeiten $P_1(Z_0; 0, T)$ sind Integrale der Form

$$\int\int_{0 \leq \tau_1 < \tau_2 \leq T} p_2(\tau_1, \tau_2)\, d\tau_1\, d\tau_2 = \tfrac{1}{2} \int_0^T\int_0^T q_2(\tau_1, \tau_2)\, d\tau_1\, d\tau_2 \tag{11}$$

auszuwerten. Falls $q_2(x, y)$ in $0 \leq x \leq T$, $0 \leq y \leq T$ in analytischer Form vorliegt oder an gewissen unten näher erläuterten Stützpunkten gegeben ist, berechnen wir dieses bestimmte Doppelintegral mit einer von LOWAN-DAVIDS-LEVENSON [6] angegebenen Methode mit Hilfe von Legendreschen Polynomen. Mit für verschiedene N_1 bzw. N_2 tabellierten Werten $a_i^{N_1}$, $x_i^{N_1}$, $i = 1, \dots, N_1$ bzw. $a_j^{N_2}$, $y_j^{N_2}$, $j = 1, \dots, N_2$ erhält man

$$\int_0^T\int_0^T q_2(x, y)\, dx\, dy = \frac{T^2}{4} \sum_{i=1}^{N_1}\sum_{j=1}^{N_2} a_i^{N_1} a_j^{N_2} q_2\left(\frac{T}{2}[x_i^{N_1} + 1], \frac{T}{2}[y_j^{N_2} + 1]\right). \tag{12}$$

Bei Benutzung von (12) müssen die Werte der Funktion an den Stellen $\frac{T}{2}[x_i^{N_1} + 1]$, $\frac{T}{2}[y_j^{N_2} + 1]$ vorliegen. Wählen wir insbesondere $N_1 \neq N_2$, so ist stets $x_i^{N_1} \neq y_j^{N_2}$, und es werden zur numerischen Berechnung keine Funktionswerte für zusammenfallende Argumente, d. h. an singulären Stellen benötigt, so daß der Übergang von p_2 zu q_2 nicht erforderlich ist. Liegt $p_2(\tau_1, \tau_2)$ an äquidistanten Stützstellen $i \cdot \Delta\tau_1$, $k \cdot \Delta\tau_2$, $\Delta\tau_j = n_j \cdot \Delta t$, $j = 1, 2$, $i = 1, \dots, K_1$, $k = 1, \dots, K_2$ berechnet oder gemessen vor, so muß zur Bestimmung des Integrals (11) ein numerisches Integrationsverfahren mit konstanter Schrittweite benutzt werden, z. B. die Simpsonsche Regel (s. ZURMÜHL [7]).

Ähnlich ist das Vorgehen im Falle von stationären stochastischen Prozessen $Z(t)$ und $\dot{Z}(t)$. Die Funktion $p_2(\tau)$ ist für $\tau \neq 0$ stetig, symmetrisch und beschränkt. Für $\tau = 0$ dagegen ist sie singulär. Mit $\tau = \tau_2 - \tau_1$ und

$$q_2(\tau) = \begin{cases} p_2(\tau) & \text{für } \tau \neq 0 \\ p_1^2 & \phantom{\text{für }} \tau = 0 \end{cases} \tag{13}$$

folgt

$$\int\int_{0 \leq \tau_1 < \tau_2 \leq T} p_2(\tau_2 - \tau_1)\, d\tau_1\, d\tau_2 = \int_{T > 0}^{T} (T - \tau)\, p_2(\tau)\, d\tau = \int_0^T (T - \tau)\, q_2(\tau)\, d\tau\,.$$

Das letzte Integral kann auf verschiedene Arten ausgewertet werden. Fallen z. B. die Stützstellen τ_i mit den Nullstellen $x_i^{N_1}$ des Legendreschen Polynoms N_1-ten Grades zusammen, folgt mit den gleichen Bezeichnungen wie oben

$$\int_0^T (T - x)\, q_2(x)\, dx \approx \frac{T}{2} \sum_{i=1}^{N_1} a_i^{N_1}\left(T - \frac{T}{2}[x_i^{N_1} + 1]\right) q_2\left(\frac{T}{2}[x_i^{N_1} + 1]\right). \tag{14}$$

Ist die Funktion $p_2(\tau)$ an einer endlichen Anzahl von äquidistanten Stützstellen $i \cdot \Delta\tau$, $i = 1, \dots, K_1$ vorgegeben läßt sich das Integral z. B. durch die Simpsonsche Regel approximieren.

5. Numerische Beispiele

Anhand von zwei einfachen Beispielen wird die praktische Nutzung obigen Vorgehens gezeigt.

a) An einem Bauwerk sind über einen längeren Zeitraum Winddruckmessungen durchgeführt worden. Die in einer fixierten Höhe zu Zeitabständen von 0,2 sec aufgenommenen Werte des Winddruckes und die daraus resultierende zeitliche Ableitung werden als Realisierungen der ergodischen stochastischen Prozesse $Z(t)$ und $\dot{Z}(t)$ angesehen. Die Auswertung hat zum Ziel, Wahrscheinlichkeiten für das Überschreiten eines vorgegebenen Niveaus Z_0 innerhalb bestimmter Zeitintervalle $[0, T)$ zu ermitteln. Wir wählen mit DE als Druckeinheit als Klassengrenzen für die Werte der Ableitungen $\bar{z}_{1,k} = y_k^{15}$ DE $\cdot$ sec^{-1}. Die Auszählung der Realisierung vom Umfang $M = 10^8$ Meßwerten ergab folgende absolute Klassenhäufigkeiten

k	1	2	3	4	5	6	7	8	9	10	11	12	13	14	15
$H_{Z_0}^M(k)$	2	5	4	3	1	2	1	3	2	1	2	1	1	0	1

Damit erhält man nach (9)

$$p_1 = 2{,}872 \cdot 10^{-6}\,\text{sec}^{-1}\,. \tag{15}$$

Zur Berechnung von $q_2(\tau)$ wird $\tau = i \cdot \Delta\tau$ mit $i = 1, \dots, 10$; $q_2(0) = p_1^2$ und $\Delta\tau = 1$ sec gewählt. Die Auszählung der Realisierungen auf gemeinsame Überschreitungen im Zeitabstand τ ergab folgende Klassenhäufigkeiten (nichtaufgeführte $H_{Z_0}^M(k, l)$ haben den Wert Null)

$$\tau = \Delta\tau: \quad H^M_{Z_0}(1, 1) = 1; \quad H^M_{Z_0}(2, 2) = 3; \quad H^M_{Z_0}(3, 3) = H^M_{Z_0}(1, 2) = H^M_{Z_0}(2, 1) = 1$$

$$\tau = 2\Delta\tau: \quad H^M_{Z_0}(1, 1) = 1; \quad H^M_{Z_0}(2, 2) = 2; \quad H^M_{Z_0}(3, 3) = H^M_{Z_0}(1, 2) = H^M_{Z_0}(3, 1) = 1$$

$$\tau = 3\Delta\tau: \quad H^M_{Z_0}(1, 1) = H^M_{Z_0}(2, 2) = H^M_{Z_0}(2, 3) = H^M_{Z_0}(2, 1) = 1$$

$$\tau = 4\Delta\tau: \quad H^M_{Z_0}(2, 2) = H^M_{Z_0}(3, 3) = H^M_{Z_0}(1, 5) = 1$$

$$\tau = 5\Delta\tau: \quad H^M_{Z_0}(2, 2) = H^M_{Z_0}(3, 3) = 1$$

$$\tau = 6\Delta\tau: \quad H^M_{Z_0}(2, 2) = 1$$

$$\tau = 7\Delta\tau: \quad H^M_{Z_0}(1, 1) = 1$$

$$\tau = 8\Delta\tau: \quad H^M_{Z_0}(2, 2) = 1$$

$$\tau = 10\Delta\tau: \quad H^M_{Z_0}(2, 2) = 1$$

Nach (13) ergibt sich für $T > 10 \cdot \Delta\tau$

$$\int_0^T (T - \tau)\, q_2(\tau)\, d\tau = 1{,}880 \cdot 10^{-8}\,\mathrm{sec}^{-2} \cdot T - 5{,}673 \cdot 10^{-8}\,\mathrm{sec}^{-1}\,.$$

Die ersten beiden Näherungen $P_{1_1}(Z_0; 0, T)$ und $P_{1_2}(Z_0; 0, T)$ der Überschreitenswahrscheinlichkeit $P_1(Z_0; 0, T)$ werden nach Friedrich-Hennig [4] bestimmt. Für verschiedene Zeitintervalle $[0, T)$ erhalten wir

T		$P_{1_1}(Z_0; 0, T)$	$P_{1_2}(Z_0; 0, T)$
1 min	$= 60\ \Delta\tau$	0,000172	0,000171
30 min	$= 1{,}8000 \cdot 10^3\ \Delta\tau$	0,005153	0,005135
1 h	$= 3{,}6000 \cdot 10^3\ \Delta\tau$	0,010286	0,010271
1 Tag	$= 8{,}6400 \cdot 10^4\ \Delta\tau$	0,219750	0,242161
3 Tage	$= 2{,}5920 \cdot 10^5\ \Delta\tau$	0,524991	0,638165
1 Woche	$= 6{,}04801 \cdot 10^5\ \Delta\tau$	0,823950	0,960577
2 Wochen	$= 1{,}2096 \cdot 10^6\ \Delta\tau$	0,969006	0,999924
1 Monat	$= 2{,}4192 \cdot 10^6\ \Delta\tau$	0,999039	1

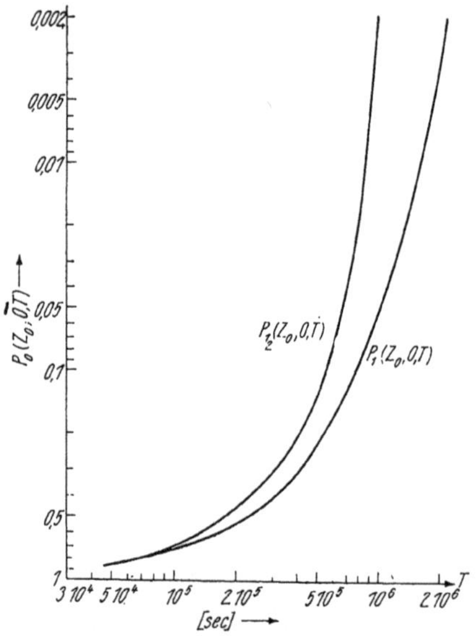

Bild 2. Erste und zweite Näherung für die Nichtüberschreitenswahrscheinlichkeit $P_0(Z_0; 0, T)$ beim Beispiel 1

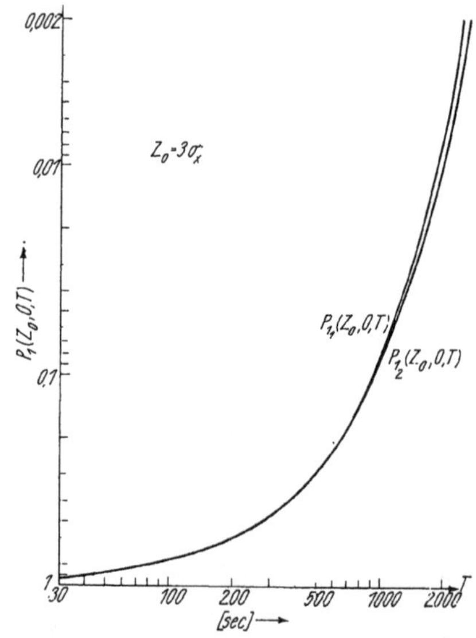

Bild 3. Erste und zweite Näherung für die Nichtüberschreitenswahrscheinlichkeit $P_0(Z_0; 0, T)$ beim Beispiel 2 mit $Z_0 = 3\sigma_Z$

Die für verschiedene T-Werte erhaltenen Kurven für $P_{0_1}(Z_0; 0, T) = 1 - P_{1_1}(Z_0; 0, T)$ und für $P_{0_2}(Z_0; 0, T) = 1 - P_{1_2}(Z_0; 0, T)$ sind in Bild 2 graphisch dargestellt.

b) Es sei $x(t)$ ein normalverteilter, stationärer stochastischer Prozeß, der durch die statistischen Parameter Mittelwert $E[x(t)] = 0$ und Korrelationsfunktion $E[x(t)\, x(t + \tau)] = K_{xx}(\tau) = e^{-\tau^2}$ bestimmt ist. Gesucht sind die Überschreitenswahrscheinlichkeiten $P_{1_1}(Z_0; 0, T)$ für ein niedriges Niveau $Z_0 = 3\sigma_x$ und ein hohes Niveau $Z_0 = 6\sigma_x$. Nach den bei Friedrich-Hennig [4] angegebenen Formeln (23) und unter Benutzung der oben beschriebenen numerischen Verfahren werden die ersten beiden Näherungen für die Überschreitenswahrscheinlichkeiten berechnet. Es ergibt sich:

$$Z_0 = 3\sigma_x: \quad P_{1_1}(Z_0; 0, T) = 1 - \exp\{-2{,}500401 \cdot 10^{-3} \cdot T\}\,,$$

$$P_{1_2}(Z_0; 0, T) = 1 - \exp\{-2{,}507259 \cdot 10^{-3} \cdot T + 8{,}3904 \cdot 10^{-8} \cdot T^2 + 4{,}611863 \cdot 10^{-6}\}\,;$$

$$Z_0 = 6\sigma_x: \quad P_{1_1}(Z_0; 0, T) = 1 - \exp\{-3{,}427948 \cdot 10^{-9} \cdot T\}\,,$$

$$P_{1_2}(Z_0; 0, T) = 1 - \exp\{-3{,}427947 \cdot 10^{-9} \cdot T + 1{,}57699 \cdot 10^{-19} \cdot T^2 - 4{,}774167 \cdot 10^{-16}\}\,.$$

Daraus erhält man für verschiedene Zeiten T folgende Werte ($\Delta\tau = 1$):

$Z_0 = 3\sigma_Z$

T	$P_{1_1}(Z_0; 0, T)$	$P_{1_2}(Z_0; 0, T)$
1 min = 60 $\Delta\tau$	0,139313	0,139403
3 min = 180 $\Delta\tau$	0,362418	0,361468
5 min = 300 $\Delta\tau$	0,527690	0,525086
10 min = 600 $\Delta\tau$	0,776924	0,771026
20 min = 1200 $\Delta\tau$	0,950237	0,944306
40 min = 2400 $\Delta\tau$	0,997524	0,996050

$Z_0 = 6\sigma_Z$

T	$P_{1_1}(Z_0; 0, T)$	$P_{1_2}(Z_0; 0, T)$
1 Jahr = $3{,}1104 \cdot 10^7\ \Delta\tau$	0,101135	0,100998
3 Jahre = $9{,}3312 \cdot 10^7\ \Delta\tau$	0,273756	0,272758
5 Jahre = $1{,}5557 \cdot 10^8\ \Delta\tau$	0,413225	0,410983
10 Jahre = $3{,}1104 \cdot 10^8\ \Delta\tau$	0,655696	0,650402
20 Jahre = $6{,}2208 \cdot 10^8\ \Delta\tau$	0,881455	0,873995
50 Jahre = $1{,}5552 \cdot 10^9\ \Delta\tau$	0,995162	0,992915

Die entsprechenden bei verschiedenen T-Werten erhaltenen Kurven für $P_{0_1}(Z_0; 0, T) = 1 - P_{1_1}(Z_0; 0, T)$ und für $P_{0_2}(Z_0; 0, T) = 1 - P_{1_2}(Z_0; 0, T)$ sind in Bild 3 und 4 graphisch dargestellt.

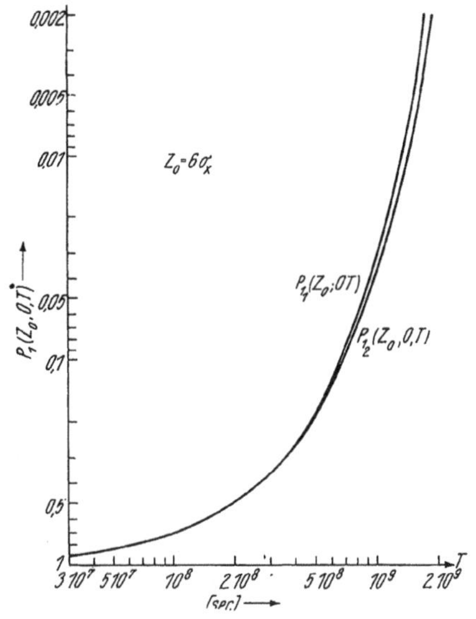

Bild 4. Erste und zweite Näherung für die Nichtüberschreitenswahrscheinlichkeit $P_0(Z_0; 0, T)$ beim Beispiel 2 mit $Z_0 = 6\sigma_Z$

Literatur

1 HEINRICH, W., HENNIG, K., Zufallsschwingungen mechanischer Systeme, Akademie-Verlag, Berlin 1977.
2 LIN, Y. K., First-Excursion Failure of Randomly Excited Structures, AIAA Journal, 8, 720–725 (1970).
3 YANG, J.-N., SHINOZUKA, M., On the First-Excursion Probability in Stationary Narnow-Band Random Vibration, Journal of Applied Mechanics, 38, 1017–1022, (1971), 39, 733–738 (1972).
4 FRIEDRICH, H., HENNIG, K., Zur Berechnung von Überschreitensversagenswahrscheinlichkeiten stochastisch beanspruchter mechanischer Systeme, ZAMM 60, 177–183 (1980).
5 SALZER, H. E., ZUCKER, R., Table of the Zeros and Weight Factors of the First Fifteen Laguerre Polynomials, Bull. amer. math. Soc., 55, 1004–1012 (1949).
6 LOWAN, A. N., DAVIDS, N., LEVENSON, A., Table of the Zeros of the Legendre Polynomials of Order 1–16 and the Weight Coefficients for Gauss' Mechanical Quadrature Formula, Bull. amer. math. Soc., 48, 739–743 (1942).
7 ZURMÜHL, R., Praktische Mathematik für Ingenieure und Physiker, Springer-Verlag, Berlin-Heidelberg-New York 1965.

Eingereicht am 9. 7. 1979, in revidierter Fassung am 16. 10. 1979

Anschrift: Dr. sc. nat. HERMANN FRIEDRICH, Prof. Dr. sc. techn. KLAUS HENNIG, ZIMM der AdW der DDR, 108 Berlin, Mohrenstraße 39, DDR

KLEINE MITTEILUNGEN

ZAMM 60, 263 – 265 (1980)

J. G. VERWER

On Generalized Runge-Kutta Methods Using an Exact Jacobian at a Non-Step Point

1. Introduction

Let

$$y' = f(y)\,, \qquad y(x_0) = y_0 \tag{1}$$

represent the initial value problem for a *stiff system* of ordinary differential equations, written in autonomous form, of which the real vector function $f(y)$ is sufficiently differentiable. An important class of numerical integration methods for this problem are formed by the so-called ROSENBROCK formulas [7]. A ROSENBROCK formula may be characterized as an explicit, one-step RUNGE-KUTTA formula of which the scalar parameters have been replaced by particular rational functions of the step-size h and the JACOBIAN matrix $J(y) = \partial f(y)/\partial y$. The aim of this replacement is to obtain attractive stability properties, such as A-stability. In the original ROSENBROCK method the treatment of each new RUNGE-KUTTA stage, which means a new f-evaluation, may require a new J-evaluation. As each J-evaluation also involves an LU-decomposition, this is not recommended. Consequently, most authors discussing ROSENBROCK type methods consider schemes which evaluate J once per integration step, viz. at the *step point* $y = y_n$ [1, 2, 4, 6, 7, 10, 11, 12]. However, if we allow *one J-evaluation per integration step*, it is also possible to investigate schemes which evaluate J at some *non-step point*, rather than at $y = y_n$. The aim is, e.g., the development of schemes giving more accuracy at the cost of the same number of operations, or the development of schemes which are more flexible with respect to the implementation of some error control mechanism.

Two first investigations in this direction have been reported by SCHOLZ, BRÄUER and THOMAS [8] and SCHOLZ [9], respectively. SCHOLZ [9] investigates a modification of the original ROSENBROCK method. He constructs a one-stage formula and several two-stage formulas, all evaluating J at a non-step point. In addition, he is able to show that his formulas satisfy the *S-stability requirements* [12].

In this note we apply the idea of SCHOLZ, of evaluating the JACOBIAN once per integration step at a non-step point, to the two-stage formula

$$y_{n+1} = y_n + \Theta_0(hJ_{n+\eta})\, hf(y_n) + \Theta_1(hJ_{n+\eta})\, hf(y_n + \\ + \Lambda(hJ_{n+\eta})\, hf(y_n)) \,. \tag{2}$$

Here $J_{n+\eta} = J(y_n + \eta h f(y_n))$, η a scalar, and Θ_0, Θ_1 and Λ are rational functions with real coefficients. Following VAN DER HOUWEN [10], we do not specify these functions beforehand and call (2) a *generalized* RUNGE-KUTTA *formula*. Class (2) contains the modified ROSENBROCK formulas constructed by SCHOLZ.

2. The local truncation error and the stability function

Let the operator equation $y_{n+1} = E_n(y_n)$ represent (2). Let y be an exact and sufficiently differentiable solution for (1). Denote

$$\theta_i = \Theta_i(0)\,,\ \theta_i' = \frac{\mathrm{d}}{\mathrm{d}z}\,\Theta(z)|_{z=0}\,,\ \ \lambda = \Lambda(0)\,,\ \lambda' = \frac{\mathrm{d}}{\mathrm{d}z}\,\Lambda(z)|_{z=0}\,, \dots$$

By means of TAYLOR's theorem for functions of several variables, and using tensor notation [3], the *local truncation error* $y(x_{n+1}) - E_n(y(x_n))$ can be expanded as

$$y(x_{n+1}) - E_n(y(x_n))$$

$$= h[1 - (\theta_0 + \theta_1)]f +$$

$$+ h^2[1/2 - (\theta_0' + \theta_1' + \theta_1\lambda)]f_jf^j +$$

$$+ h^3\left[\frac{1}{6} - \left(\frac{1}{2}\theta_0'' + \theta_1\lambda' + \theta_1'\lambda + \frac{1}{2}\theta_1''\right)\right]f_jf^j_kf^k +$$

$$+ h^3\left[\frac{1}{6} - \left(\theta_0'\eta + \frac{1}{2}\theta_1\lambda^2 + \theta_1'\eta\right)\right]f_{jk}f^jf^k +$$

$$+ h^4\left[\frac{1}{24} - \left(\frac{1}{6}\theta_0''' + \frac{1}{2}\theta_1\lambda'' + \theta_1'\lambda' + \frac{1}{2}\theta_1''\lambda + \right.\right.$$

$$\left.\left. + \frac{1}{6}\theta_1'''\right)\right]f_jf^j_kf^k_lf^l +$$

$$+ h^4\left[\frac{1}{24} - \left(\frac{1}{2}\theta_0''\eta + \eta\theta_1\lambda' + \frac{1}{2}\theta_1'\lambda^2 + \frac{1}{2}\theta_1''\eta\right)\right]f_jf^j_{kl}f^kf^l +$$

$$+ h^4\left[\frac{3}{24} - \left(\frac{1}{2}\theta_0''\eta + \lambda\lambda'\theta_1 + \lambda\eta\theta_1' + \frac{1}{2}\theta_1''\eta\right)\right]f_{jk}f^j_lf^kf^l +$$

$$+ h^4\left[\frac{1}{24} - \left(\frac{1}{2}\theta_0'\eta^2 + \frac{1}{6}\theta_1\lambda^3 + \frac{1}{2}\theta_1'\eta^2\right)\right]f_{jkl}f^jf^kf^l +$$

$$+ h^5\left[\frac{1}{120} - \left(\frac{1}{24}\theta_0'''' + \frac{1}{6}\theta_1\lambda''' + \frac{1}{2}\theta_1'\lambda'' + \frac{1}{2}\theta_1''\lambda' + \right.\right.$$

$$\left.\left. + \frac{1}{6}\theta_1'''\lambda + \frac{1}{24}\theta_1''''\right)\right]f_jf^j_kf^k_lf^l_mf^m +$$

$$+ h^5\left[\frac{3}{120} - \left(\frac{1}{6}\theta_0'''\eta + \frac{1}{2}\theta_1\lambda''\eta + \theta_1'\lambda\lambda' + \frac{1}{2}\theta_1''\eta\lambda + \right.\right.$$

$$\left.\left. + \frac{1}{6}\theta_1'''\eta\right)\right]f_jf^j_{kl}f^k_mf^lf^m +$$

$$+ h^5\left[\frac{1}{120} - \left(\frac{1}{6}\theta_0'''\eta + \frac{1}{2}\theta_1\lambda''\eta + \theta_1'\eta\lambda' + \frac{1}{4}\theta_1''\lambda^2 + \right.\right. \tag{3}$$

$$\left.\left. + \frac{1}{6}\theta_1'''\eta\right)\right]f_jf^j_kf^k_{lm}f^lf^m +$$

$$+ h^5\left[\frac{1}{120} - \left(\frac{1}{4}\theta_0''\eta^2 + \frac{1}{2}\theta_1\lambda'\eta^2 + \frac{1}{6}\theta_1'\lambda^3 + \right.\right.$$

$$\left.\left. + \frac{1}{4}\theta_1''\eta^2\right)\right]f_jf^j_{klm}f^kf^lf^m +$$

$$+ h^5\left[\frac{3}{120} - \frac{1}{2}\theta_1\lambda'^2\right]f_{jk}f^j_lf^k_mf^lf^m +$$

$$+ h^5\left[\frac{4}{120} - \left(\frac{1}{6}\theta_0'''\eta + \frac{1}{2}\theta_1\lambda\lambda'' + \theta_1'\lambda'\eta + \frac{1}{2}\theta_1''\eta\lambda + \right.\right.$$

$$\left.\left. + \frac{1}{6}\theta_1'''\eta\right)\right]f_{jk}f^j_lf^l_mf^kf^m +$$

$$+ h^5\left[\frac{4}{120} - \left(\frac{1}{2}\theta_0''\eta^2 + \lambda\lambda'\eta\theta_1 + \frac{1}{2}\theta_1'\eta\lambda^2 + \frac{1}{2}\theta_1''\eta^2\right)\right]f_{jk}f^j_{lm}f^kf^lf^m +$$

$$+ h^5\left[\frac{6}{120} - \left(\frac{1}{4}\theta_0''\eta^2 + \frac{1}{2}\theta_1\lambda^2\lambda' + \frac{1}{2}\theta_1'\lambda\eta^2 + \right.\right.$$

$$\left.\left. + \frac{1}{4}\theta_1''\eta^2\right)\right]f_{jkl}f^j_mf^kf^lf^m +$$

$$+ h^5\left[\frac{1}{120} - \left(\frac{1}{6}\theta_0'\eta^3 + \frac{1}{24}\theta_1\lambda^4 + \frac{1}{6}\theta_1'\eta^3\right)\right]f_{jklm}f^jf^kf^lf^m + O(h^6)\,.$$

In the sequel, the integer p will stand for *the order of consistency*, i.e. p is the largest integer satisfying $y(x_{n+1}) - E_n(y(x_n)) = O(h^{p+1})$, $h \to 0$. Using the theory of BUTCHER series, VAN KAMPEN [11] has shown that for an m-stage method of type (2), with $\eta = 0$, the other p cannot exceed $2m$. For the case $m \leqq 2$ this result can be verified by inspection of (3). If $\eta = 0$ and $m = 1$, i.e. $\theta_1 = 0$, the differential $f_{jk}f^jf^k$ is not present and if $\eta = 0$ and $m = 2$, we see that $f_{jk}f^j_{lm}f^kf^lf^m$ is missing. An interesting question is now, whether by the introduction of the parameter η an increase of the order can be realized. It turns out that for the one-stage formula $p = 3$ can be obtained (see [9] and section 3), whereas for the two-stage formula we always have $p \leqq 4$ (see section 4).

When applied to the *stability test-model* $y' = \delta y$, $\delta \in \mathbb{C}$, (2) will result into the scalar relation $y_{n+1} = R(z)\, y_n$, $z = h\delta$, where

$$R(z) = 1 + z\Theta_0(z) + z\Theta_1(z) + z^2\Theta_1(z)\, \Lambda(z) \,. \tag{4}$$

The stability properties of (2), such as A-stability, are determined by the so-called *stability function* R (see e.g. [5]). In the remaining part of the note we shall make frequent use of the concept of stability function, Herewith we use the following definitions (cf. [5]): R is said to be (a) *A-acceptable*, if $|R(z)| < 1$ whenever Re $(z) < 0$; (b) *strongly A-acceptable*, if it is A-acceptable and satis fies lim $|\text{Re}\,(z)| < 1$ as Re $(z) \to -\infty$; (c) *L-acceptable*, if it is A-acceptable and satisfies lim $R(z) = 0$ as Re $(z) \to -\infty$. Further, we let q denote *the order of consistency of R*, i.e. q is the largest integer satisfying $R(z) - e^z = O(z^{q+1})$, $z \to 0$.

3. One-stage formulas of order three

In this section we concentrate on the class of formulas $(\Theta_1 = 0$ in (2))

$$y_{n+1} = y_n + \Theta_0(hJ_{n+\eta})\, hf(y_n) \,. \tag{5}$$

Theorem 1: *Let I denote the unit matrix. Let R represent some adaptive stability function or order q. The formula*

$$y_{n+1} = y_n + (hJ_{n+\eta})^{-1}\,(R(hJ_{n+\eta}) - I)\, hf(y_n) \tag{6}$$

is of order $p = 3$ if $\eta = \frac{1}{3}$ and $q \geqq 3$. If η is arbitrary and $q \geqq 2$, we always have $p = 2$.

The proof follows from a simple calculation with (3). By pursuing this calculation somewhat further, it is also easy to see that we always have $p \leqq 3$. The applicability of this theorem lies in the *adaptivity* of the stability function R. In fact, (6), where $\eta = \frac{1}{3}$ and $q \geqq 3$, represents a large class of third order integration formulas of which, to a certain extent, the stability behaviour can be adapted to the problem under consideration (cf. [10]). For example, we can choose the third order stability function

$$R(z) = \frac{1 + (1 + \alpha)\, z + \left(\frac{1}{3} + \frac{1}{2}\alpha\right) z^2}{1 + \alpha z - \left(\frac{1}{6} + \frac{1}{2}\alpha\right) z^2} \tag{7}$$

of Liniger and Willoughby (see [10], p. 79), where the free parameter α can be used for exponential fitting. Function (7) is A-acceptable if $\alpha \leqq -\frac{1}{2}$ and L-acceptable if $\alpha = -\frac{2}{3}$. Observe that if the stability function R in (6) is strongly A-acceptable, the method is also S-stable [12]. Another member of class (6) is defined by the third order function given by Scholz [9], viz.

$$R(z) = \frac{1 - \frac{1}{3}\sqrt{3}\, z - \left(\frac{1}{6} + \frac{1}{6}\sqrt{3}\right) z^2}{\left(1 - \left(\frac{1}{2} + \frac{1}{6}\sqrt{3}\right) z\right)^2} \tag{8}$$

which is strongly A-acceptable. An attractive property of (8), with respect to the aspect of computer implementation (see [9]), is that the denominator is factorized into two equal linear terms.

4. Two-stage formulas of order four

Following [10, section 2.7.8], we express the derivatives of Θ_0, at $z = 0$, into the derivatives of Θ_1, Λ and R:

$$\begin{aligned} \theta_0 &= R'(0) - \theta_1\,, \quad \theta_0' = \frac{1}{2} R''(0) - \theta_1' - \theta_1\lambda\,, \\ \theta_0'' &= \frac{1}{3} R'''(0) - \theta_1'' - 2\theta_1'\lambda - 2\theta_1\lambda'\,, \\ \theta_0''' &= \frac{1}{4} R''''(0) - \theta_1''' - 3\theta_1''\lambda - 6\theta_1'\lambda' - 3\theta_1\lambda''\,. \end{aligned} \tag{9}$$

Assuming that R is a 4-th order consistent approximation to e^z and substitution of (9) into the $(p \leqq 4)$-expressions of expansion (3), yields the remaining conditions for 4-th order accuracy:

$$\begin{aligned} &\eta\left(\frac{1}{2} - \theta_1\lambda\right) = \frac{1}{6} - \frac{1}{2}\theta_1\lambda^2\,, \qquad \eta^2\left(\frac{1}{2} - \theta_1\lambda\right) = \frac{1}{12} - \frac{1}{3}\theta_1\lambda^3\,, \\ &\eta\left(\frac{1}{6} - \theta_1'\lambda\right) = \frac{1}{24} - \frac{1}{2}\theta_1'\lambda^2\,, \qquad \eta\left(\frac{1}{6} - \theta_1\lambda'\right) = \frac{3}{24} - \lambda\lambda'\theta_1\,. \end{aligned} \tag{10}$$

If the new parameter $\eta = 0$, the remaining parameters are: $\theta_1 = \frac{16}{27}$, $\lambda = \frac{3}{4}$, $\theta_1' = \frac{4}{27}$, $\lambda' = \frac{9}{32}$. First we prove the following negative result:

Theorem 2: *The introduction of the parameter η into the 2-stage generalized* Runge-Kutta *method does not result into an increase of the order of consistency. The maximal order remains $p = 4$.*

Proof: We only need to show that $p = 5$ is impossible. Let $\xi = \theta_1\lambda$. The elimination of η between the first two equations of (10) yields

$$(6\lambda^2)\,\xi^2 + (12\lambda - 12\lambda^2 - 6)\,\xi + 1 = 0\,. \tag{11}$$

The last term in (3) yields the order relation $\eta^3\left(\frac{1}{2} - \xi\right) = \frac{1}{20} - \frac{1}{4}\xi\lambda^3$. By again eliminating η, we then find

$$(30\lambda^3)\,\xi^2 + (15\lambda + 20\lambda^2 - 45\lambda^3 - 18)\,\xi + 4 = 0\,. \tag{12}$$

A simple calculation reveals that (11) and (12) are incompatible.

The solution of system (10) can be written as the one-parameter solution

$$\begin{aligned} \theta_1 &= \left(1 - 2\lambda + 2\lambda^2 + \sqrt{(1 - 2\lambda + 2\lambda^2)^2 - \frac{2}{3}\lambda^2}\right)\Big/\zeta\,, \\ \eta &= \left(\frac{1}{6} - \frac{1}{2}\theta_1\lambda^2\right)\Big/\left(\frac{1}{2} - \theta_1\lambda\right), \\ \theta_1' &= \left(\frac{1}{24} - \frac{1}{6}\eta\right)\Big/\left(\frac{1}{2}\lambda^2 - \eta\lambda\right), \\ \gamma' &= \left(\frac{3}{24} - \frac{1}{6}\eta\right)\Big/(\lambda\theta_1 - \eta\theta_1)\,, \end{aligned} \tag{13}$$

where $\zeta = \frac{81}{64}$ if $\lambda = \frac{3}{4}$ and $\zeta = 2\lambda^3$ if $\lambda \neq \frac{3}{4}$. The free parameter λ can take all values except zero. We are thus led to the following theorem:

Theorem 3: *Let R be an adaptive stability function of order $q \geqq 4$. Equations (9) and (13) then define a four-parameter class of generalized Runge-Kutta formulas of order $p = 4$. The free parameters are θ_1', θ_1'', λ'' and $\lambda (\lambda \neq 0)$.*

Of interest, with respect to the stability behaviour, is the freedom which is left in the choice of R and the functions Θ_0, Θ_1 and Λ. A nice example is the fourth order modified Rosenbrock formula given by Scholz [9]. The stability function of that formula, which also appears in another type of Rosenbrock formula developed by Kaps [4], is A-acceptable and factorized. Further, the functions Θ_0, Θ_1 and Λ are such that the formula is also internally S-stable (cf. [12]). Another example of a formula belonging to class (9), (13) is the L-stable formula discussed by van Kampen [11]. Here the parameter $\eta = 0$.

5. Some final remarks

From the foregoing it shall be clear that if we consider 2-stage formulas of order $p = 3$, we have a considerable degree of freedom with respect to the parameter choice. One possibility is to construct a scheme with a build in error control. It is an easy task to show that, thanks to the introduction of the parameter η, one can construct 2-stage formulas of order $p = 3$ such that $y_{n+1} - [y_n + \Lambda(hJ_{n+\eta})\, f(y_n)]$ provides an estimate of $y'''(x_n)$. Another conservative error estimator, which can be used for every 1-stage formula (6), is provided by the expression $J_{n+1+\eta}f(y_{n+1}) - J_{n+\eta}f(y_n)$, where $\eta = \frac{1}{3}$. This expression also estimates $y'''(x_n)$.

For a 2-stage formula order 4 can already be achieved with four degrees of freedom. Therefore it is likely that $p = 5$ can be obtained if we perform a third f-evaluation. Along the lines of section 4 it is then possible to develop a family of 5-th order, 3-stage formulas with an adaptive stability function. In this connection, the stability functions discussed by Kaps [4] are of interest.

One might also think of generalized RUNGE-KUTTA schemes which evaluate the JACOBIan at some weighted sum $\alpha y_n + \beta y_{n-1} + \ldots$ Here we avoid the occurrence of the *explicit* expression $y_n + \eta h f(y_n)$ in the J-evaluation. For stiff, highly non-linear problems, that is problems with a strongly varying JACOBIan $J(y)$, we then expect a somewhat better stability-accuracy behaviour.

References

1 CALAHAN, D. A., A stable accurate method of numerical integration for non-linear systems, Proc. IEEE 56, 744–747 (1968).
2 FRIEDLI, A., Verallgemeinerte RUNGE-KUTTA-Verfahren zur Lösung steifer Differentialgleichungssysteme, Berlin-Heidelberg-New York, Springer-Verlag, Lecture Notes in Mathematics 631, 1978, 35–50.
3 HENRICI, P., Discrete Variable Methods in Ordinary Differential Equations, John Wiley & Sons, New York 1962.
4 KAPS, P., Modifizierte ROSENBROCK-Methoden der Ordnung 4, 5 und 6 zur numerischen Integration steifer Differentialgleichungen, Dissertation, Leopold Franzens Universität, Innsbruck 1977.
5 LAMBERT, J. D., Computational Methods in Ordinary Differential Equations, John Wiley & Sons, London 1973.
6 NØRSETT, S. P.; WOLFBRANDT, A., Order conditions for ROSENBROCK type methods, Numer. Math. 32, 1–15 (1979).
7 ROSENBROCK, H. H., Some general implicit processes for the numerical solution of differential equations, Computer J. 5, 329–330 (1963).
8 SCHOLZ, S.; BRÄUER, K.; THOMAS, S., Ein A-stabiles einstufiges ROSENBROCK-Verfahren dritter Ordnung, Beiträge Numer. Math. 5, 191–199 (1976).
9 SCHOLZ, S., S-stabile modifizierte ROSENBROCK-Verfahren 3. und 4. Ordnung, Sektion Mathematik, Technische Universität Dresden 1978.
10 VAN DER HOUWEN, P. J., Construction of Integration Formulas for Initial Value Problems, North-Holland Publishing Company, Amsterdam 1977.
11 VAN KAMPEN, S. P. N., Een 4e orde gegeneraliseerde RUNGE-KUTTA methode, Report NN 2/74, Mathematisch Centrum, Amsterdam (Dutch) 1974.
12 VERWER, J. G., S-stability properties for generalized RUNGE-KUTTA methods, Numer. Math. 27, 359–370 (1977).

Eingereicht: 9. 7. 1979

Anschrift: Dr. J. G. VERWER, Mathematisch Centrum, 2e Boerhaavestraat 49, 1091 AL Amsterdam, The Netherlands

ZAMM 60, 265–266 (1980)

G. U. SCHUBERT

Inversion der hyperbolischen Kepler-Gleichung mittels Laplace-Transformation

Eine KEPLER-*Hyperbel* (Achsen a und b) wird durch einen zeitabhängigen Parameter u beschrieben (Vergl. [1]):

$$\vec{x} = (a \cosh u - a\varepsilon, -b \sinh u, 0)\,, \tag{1}$$

$$\varepsilon \sinh u - u = \tau\,, \tag{2}$$

$$\tau := t \cdot 2\Omega(\varepsilon^2 - 1)^{3/2} p^{-2}\,, \tag{3}$$

$a = p(\varepsilon^2 - 1)^{-1}$, $b = p(\varepsilon^2 - 1)^{-1/2}$, Ω die Flächengeschwindigkeit.

In Lehrbüchern der Mechanik steht der Zusammenhang mit den physikalischen Größen. Hält man p und Ω fest und läßt $\varepsilon \to 1$ gehen, dann wird aus (2) eine kubische Gleichung für die Variable

$$\eta := [u(\varepsilon^2 - 1)^{-1/2}]_{\varepsilon \to 1}$$

so daß wir stets $\varepsilon > 1$ annehmen dürfen. Auch können wir uns auf $\tau \geqq 0$ beschränken. Unter Beachtung der Abbildungseigenschaften von (2) wird die LAPLACE-*Transformierte*

$$\mathcal{L}(u) =: U(s) = s^{-1} \int_0^\infty \exp(-s \cdot \tau(u))\, du\,, \quad \mathrm{Re}\,(s) \geqq \alpha > 0\,. \tag{4}$$

Diese Beziehung gilt allgemein für eine Klasse von Abbildungen. Speziell mit (2) ergibt sich

$$U(s) = s^{-1} \int_0^\infty \exp(-\varepsilon s \cdot \sinh u + su)\, du\,. \tag{5}$$

Hierfür kann man schreiben ([2], S. 117)

$$U(s) = \pi[\boldsymbol{J}_{-1}(\varepsilon s) - \boldsymbol{J}_{-1}(\varepsilon s)]\,[s \cdot \sin(\pi s)]^{-1} \tag{6}$$

oder auch ([2], S. 66 und 118)

$$U(s) = \pi s^{-1}\{\cot(\pi s)\,[\boldsymbol{J}_s(\varepsilon s) - \boldsymbol{J}_s(\varepsilon s)] - Y_s(\varepsilon s) - \boldsymbol{E}_s(\varepsilon s)\}\,; \tag{7}$$

J BESSEL-Funktion, Y NEUMANN-Funktion,
$\boldsymbol{J}$ ANGER-Funktion, $\boldsymbol{E}$ WEBER-Funktion.

Um den Integrationsweg in der Umkehrformel

$$u(\tau) = (2\pi i)^{-1} \int_{\alpha - i\infty}^{\alpha + i\infty} \exp(\tau s)\, U(s)\, ds \tag{8}$$

zweckmäßig deformieren zu können, betrachten wir die Entwicklungen ([2], S. 66 u. 117):

$$\boldsymbol{J}_{-1}(\varepsilon s) = \exp(-s \log(\varepsilon s/2)) \sum_{m=0}^{\infty} (i\varepsilon s/2)^{2m}\, [m!\,\Gamma(m + 1 - s)]^{-1}, \tag{9}$$

$$\begin{aligned} \boldsymbol{J}_{-1}(\varepsilon s) = \sum_{m=0}^{\infty} (i\varepsilon s/2)^{2m}\, \{&\cos(\pi s/2)\,[\Gamma(m + 1 - s/2) \times \\ &\times \Gamma(m + 1 + s/2)]^{-1} - (\varepsilon s/2) \sin(\pi s/2) \times \\ &\times [\Gamma(m + 3/2 - s/2)\, \Gamma(m + 3/2 + s/2)]^{-1}\}\,. \end{aligned} \tag{10}$$

$\boldsymbol{J}_{-s}(\varepsilon s)$ ist eine ganze Funktion. $\boldsymbol{J}_{-s}(\varepsilon s)$ muß auf ein Blatt eingeschränkt werden:

$$\begin{aligned} u(\tau) = (2i)^{-1} \int_{-\infty}^{0+} \exp(\tau s)\, \boldsymbol{J}_{-s}(\varepsilon s)\, [s \cdot \sin(\pi s)]^{-1}\, ds - \\ - \pi \sum_{n=0}^{\infty} \mathrm{Res}\left(\exp(\tau s)\, \boldsymbol{J}_{-s}(\varepsilon s)\, [s \cdot \sin(\pi s)]^{-1}\right)_{s=-n} \end{aligned} \tag{11}$$

Standard-Rechenmethoden liefern

$$\begin{aligned} u(\tau) = -\gamma - \log(\varepsilon/2) + \sum_{n=1}^{\infty} n^{-1} I_n(\varepsilon n) \exp(-n\tau) + \\ + \int_0^\infty \exp(-\tau\chi)\,(\log\chi) \cdot [-\tau I_\chi(\varepsilon\chi) + dI_\chi(\varepsilon\chi)/d\chi]\, d\chi\,, \end{aligned} \tag{12}$$

γ die EULER-MASCHERONI-Konstante.

Das Integral in (12) ist vom LAPLACEschen Typ. Man entwickelt:

$$\begin{aligned} I_\chi(\varepsilon\chi) = \\ = \sum_{m,\mu,\nu} (i\varepsilon/2)^{2m}\, [m!\,\mu!\,\nu!]^{-1}\, \chi^{2m+\mu+\nu} \times \\ \times [[\log(\chi\varepsilon/2)]^\mu [d^\nu(\Gamma(m + 1 + \chi))^{-1} | d\chi^\nu]_{\chi=0}\,. \end{aligned} \tag{13}$$

Die benötigten Werte für die Ableitungen der Γ-Funktion sind über die Ψ-Funktion und ihre Eigenschaften ([2], S. 15) zu berechnen. Man erhält letztlich LAPLACE-Integrale vom Typ

$$I(\tau, q, l) := \int_0^\infty \exp(-\tau\chi)\, \chi^q (\log\chi)^l\, d\chi \quad q > -1,\ l = 0, 1, 2, \ldots \tag{14}$$

In der Literatur (z. B. [3]) habe ich außer den Fällen $l = 0$ nur die Fälle $l = 1, 2$ gefunden. Daher sei die allgemeine Formel mitgeteilt:

$$I(\tau, q, l) = [d^l I(\tau, q + z, 0)/dz^l]_{z=0}\,, \tag{15}$$

Dafür läßt sich formal schreiben:

$$I(\tau, q, l) = \tau^{-q-1}[(d/dz - \log\tau)^l\, \Gamma(q + 1 + z)]_{z=0}\,. \tag{16}$$

Umgekehrt kann man durch Bildung geeigneter Linearkombinationen von $I(\tau, q, 0)$ bis $I(\tau, q, l)$ Rücktransformation zu $\tau^{-1-q}(\log\tau)^l$ finden, z. B.

$$\tau^{-1}(\log\tau)^3 = \mathcal{L}(-2\zeta(2) + \chi^2/2(\gamma + \log\chi) - (\gamma + \log\chi)^3)\,. \tag{17}$$

Doch konnte ich hier im Gegensatz zu (16) keine einfache Form finden. Trägt man die mittels (13) und (16) erhaltenen Integrale in (12) ein, dann erhält man eine asymptotische Reihe, deren Anfang so lautet:

$$\begin{aligned} u(\tau) \sim (1 + \tau^{-1} + \tau^{-2}) \log(2\tau/\varepsilon) - \\ - (2\tau^2)^{-1}[\log(2\tau/\varepsilon) - \varepsilon^2/2] + O(\tau^{-3}(\log\tau)^3)\,. \end{aligned} \tag{18}$$

Die Regel, daß man $U(s)$ für $\mathrm{Re}\,(s) \gg 1$ zu entwickeln hat, um $u(\tau)$ für $\tau \ll 1$ zu erhalten, bringt hier nichts, was über die sofort herleitbare ([4], S. 71) TAYLOR-Reihe hinausgeht. Bei der

Verallgemeinerung des Verfahrens auf $\tau = f(u)$ muß dies nicht unbedingt der Fall sein, so daß die LAPLACE-Transformation auch für $\tau \ll 1$ weiterhilft.

Literatur

1 CHARLIER, C. L., Die Mechanik des Himmels, Band 1, Leipzig 1902, S. 192.
2 MAGNUS, W.; OBERHETTINGER, F.; SONI, R. P., Special Functions of Mathematical Physics, 3. Aufl. Berlin u.s.w. 1966.
3 OBERHETTINGER, F.; BADII, L., Tables of LAPLACE Transforms, Berlin u.s.w. 1973.
4 STUMPFF, K., Himmelsmechanik, Band III, Berlin 1974.

Eingereicht: 18. 9. 1979

Anschrift: Prof. Dr. G. U. SCHUBERT, Institut für Physik der Universität Mainz, Postfach 3980, D-6500 Mainz, BRD

ZAMM 60, 266 – 268 (1980)

J. SCHNEIDER

Der Satz von Gerschgorin für gedämpfte Schwingungen

1. Einführung

Im Eigenwertproblem

$$\{M\lambda^2 + D\lambda + C\}\, x = o \tag{1}$$

mit quadratischen Matrizen der Ordnung n sei die Steifigkeitsmatrix C normal bezüglich der positiv definiten Massenmatrix M, d. h. es gelte die Vertauschbarkeit

$$C^*M^{-1}C = CM^{-1}C^*\,, \tag{2}$$

die im Falle hermitescher Matrizen trivialerweise erfüllt ist. Dann gibt es für M und C eine unitär kongruente Simultantransformation auf Diagonalform

$$Y^*CY = \Omega\,, \tag{3}$$

$$Y^*MY = I\,, \tag{4}$$

bei der

$$\Omega := \operatorname{diag}\langle\omega_j^2\rangle\,, \qquad j = 1, 2, \dots, n \tag{5}$$

die i. A. komplexen, bei hermiteschen Matrizen M, C aber reellen (und für positiv definites bzw. semidefinites C sogar positiven bzw. nichtnegativen) Eigenwerte ω_j^2 des Paares C, M und damit die Eigenfrequenzquadrate des ungedämpften Systems enthält, während in der Modalmatrix

$$Y := \{y_1, y_2, \dots, y_n\} \tag{6}$$

die zugehörigen Eigenvektoren stehen. Mit

$$x =: Yz\,, \tag{7}$$

$$\hat{D} := (\hat{d}_{ij}) := -Y^*DY \tag{8}$$

läßt sich dann (1) in die spezielle Form

$$\{I\lambda^2 - \hat{D}\lambda + \Omega\}\, z = o \tag{9}$$

verwandeln, wobei die Transformation (8) nicht explizit durchgeführt werden muß, denn läßt man bei der Bestimmung von Ω z. B. durch JACOBI-Rotation die Dämpfungsmatrix mitlaufen, so ergibt sich $-\hat{D}$ automatisch, sobald M und C Diagonalform angenommen haben. Waren alle Matrizen hermitesch, so bleiben sie dies bei der Transformation; war überdies die Vertauschbarkeit

$$C^*M^{-1}D = DM^{-1}C^* \tag{10}$$

erfüllt (das Tripel M, D, C also nach [1] parameternormal), so besitzt auch $\hat{D}$ Diagonalform, und (9) zerfällt in n entkoppelte quadratische Gleichungen in λ [2].

2. Der Satz von GERSCHGORIN

Nun fassen wir (9) als zweiparametriges Eigenwertproblem

$$A(\lambda)\, z = \Lambda z \tag{11}$$

mit

$$A(\lambda) := (a_{ij}) := \lambda\hat{D} - \Omega \tag{12}$$

auf, vermerken als Nebenbedingung

$$\Lambda \overset{!}{=} \lambda^2 \tag{13}$$

und können den Satz von GERSCHGORIN formulieren:

In oder auf dem Rande eines jeden der Kreisgebiete mit den Radien

$$r_{iz} := \sum_{\substack{j=1\\ j\neq i}}^{n} |a_{ij}| = |\lambda| \sum_{\substack{j=1\\ j\neq i}}^{n} |\hat{d}_{ij}| \tag{14}$$

und

$$r_{is} := \sum_{\substack{j=1\\ j\neq i}}^{n} |a_{ji}| = |\lambda| \sum_{\substack{j=1\\ j\neq i}}^{n} |\hat{d}_{ji}| \tag{15}$$

um den Mittelpunkt

$$a_{ii} = \lambda\hat{d}_{ii} - \omega_i^2 \tag{16}$$

der komplexen Zahlenebene liegt mindestens ein Eigenwert Λ der Aufgabe (11):

$$|\Lambda_i - a_{ii}| \leqq \begin{cases} r_{iz} \\ r_{is} \end{cases}, \qquad i = 1, 2, \dots, n\,, \tag{17}$$

oder ausführlich:

$$|\lambda_i^2 - \lambda_i\hat{d}_{ii} + \omega_i^2| \leqq |\lambda_i|\, R_i^2\,, \qquad i = 1, 2, \dots, n \tag{18}$$

mit

$$R_i^2 := \sum_{\substack{j=1\\ j\neq i}}^{n} \begin{cases} |\hat{d}_{ij}| \\ |\hat{d}_{ji}| \end{cases}. \tag{19}$$

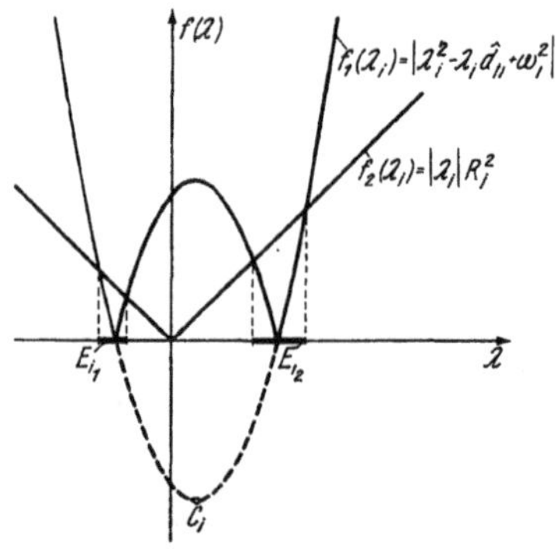

Bild 1. Einschließungsbereiche E_{i_1} und E_{i_2} für den reellen Eigenwert λ_i

3. Reelle Eigenwerte

Für hermitesche Matrizen und reelle Eigenwerte λ_i zeigt Bild 1 die nach (18) gewonnene Einschließung. Sie umfaßt all jene Bereiche, in denen die Betragskurve der mit ihrem Scheitel in den Punkt

$$C_i(\hat{d}_{ii}/2,\; \omega_i^2 - \hat{d}_{ii}^2/4) \tag{20}$$

verschobenen und nach oben geöffneten Normalparabel der linken Ungleichungsseite unterhalb der Betragsgeraden auf der rechten Seite der Ungleichung bleibt. Deren Steigung ist dem nach der Diagonalisierung von M und C in D außerhalb der Hauptdiagonalen verbleibenden Rest R_i^2 proportional. Ist dieser groß, entsteht ein zusammenhängender Einschließungsbereich, der für

$$(\hat{d}_{ii} \pm R_i^2)^2 > 4\omega_i^2 \tag{21}$$

in zwei getrennte Bereiche zerfällt und bei $R_i^2 = 0$ — d. h. bei Erfüllung von (10) — zu den exakten Eigenwerten schrumpft. Weiter zeigt man leicht, daß nur für

$$\hat{d}_{ii} < -R_i^2 \tag{22}$$

der gesamte Einschließungsbereich negativ und damit die i-te Eigenbewegung stabil ist.

4. Komplexe Eigenwerte

Zur Bestimmung komplexer Eigenwertschranken ist es möglich, λ_i in (18) als

$$\lambda_i =: r\, e^{i\varphi} \tag{23}$$

darzustellen, dann mit φ die komplexe Zahlenebene abzutasten und die zugehörigen Radiusvektoren der Einschließungsgrenzen zu berechnen. Da hierbei jedoch für jedes φ eine Gleichung vierten Grades gelöst werden muß, wollen wir einen anderen Weg gehen und mit der Summe

$$S_i^2 := R_i^2 + |\hat{d}_{ii}| \tag{24}$$

sämtlicher Zeilen- bzw. Spaltenelemente von $\hat{\boldsymbol{D}}$ einschließlich des Hauptdiagonalelementes eine erneute Abschätzung

$$|\lambda_i^2 + \omega_i^2| \leqq |\lambda_i|\, S_i^2 \tag{25}$$

vornehmen, wodurch wir mit (23) zu

$$r^4 - r^2(S_i^2 - 2\omega_i^2 \cos 2\varphi) + \omega_i^4 \leqq 0 \tag{26}$$

gelangen. Hier steht das Gleichheitszeichen für die Ränder des Einschließungsgebietes, deren Entfernungen r_i vom Ursprung damit für jeden Winkel φ aus einer biquadratischen Gleichung berechnet werden können:

$$r_{1,2}^2 = \frac{(S_i^2 - 2\omega_i^2 \cos 2\varphi)}{2} \pm \sqrt{\frac{(S_i^2 - 2\omega_i^2 \cos 2\varphi)^2}{4} - \omega_i^4}\,. \tag{27}$$

Die so gewonnenen Bereiche liegen symmetrisch zu beiden Achsen; zwei positive r_i existieren für

$$\cos 2\varphi < \frac{S_i^2}{2\omega_i^2} - 1 \tag{28}$$

und fallen am Rande des Gebiets bei

$$\cos 2\varphi = \frac{S_i^2}{2\omega_i^2} - 1 \tag{29}$$

mit

$$r^2 = \omega_i^2 \tag{30}$$

zusammen; für

$$S_i^2 > 4\omega_i^2 \tag{31}$$

umschließt jedoch nur noch ein zusammenhängendes Ringgebiet den Ursprung.

Natürlich muß jede Einschließung je einmal mit S_i^2 (vgl. (24)) als Zeilen- und Spaltensumme von $\hat{\boldsymbol{D}}$ vorgenommen werden; die Eigenwerte liegen dann im Durchschnitt beider auf diese Weise gewonnenen Gebiete. Darüber hinaus gilt auch hier der Satz, daß in ϱ Gebieten, die ein zusammenhängendes Teilgebiet bilden und alle anderen Gebiete außerhalb lassen, genau ϱ Eigenwerte liegen.

5. Beispiel

Für die Eigenwerte der Aufgabe

$$\left\{\lambda^2 \begin{bmatrix} 1 & 0 & 0 \\ 0 & 1 & 0 \\ 0 & 0 & 1 \end{bmatrix} + \lambda \begin{bmatrix} 2{,}9 & 0{,}0 & 2{,}0 \\ 0{,}0 & 1{,}0 & 0{,}0 \\ 2{,}0 & 0{,}0 & 4{,}0 \end{bmatrix} + \begin{bmatrix} 1{,}0 & -1{,}0 & 0{,}0 \\ -1{,}0 & 3{,}0 & 2{,}0 \\ 0{,}0 & 2{,}0 & 2{,}0 \end{bmatrix}\right\} \boldsymbol{x} = \boldsymbol{o} \tag{32}$$

sei die Einschließung nach Gerschgorin gesucht.

Die Transformationen (3) und (8) liefern

$$\boldsymbol{\Omega} = \begin{bmatrix} 0 & 0 & 0 \\ 0 & 1 & 0 \\ 0 & 0 & 4 \end{bmatrix}; \quad \hat{\boldsymbol{D}} = \begin{bmatrix} -0{,}9750 & 0{,}0408 & -0{,}0144 \\ 0{,}0408 & -3{,}9333 & -0{,}0236 \\ -0{,}0144 & -0{,}0236 & -0{,}9917 \end{bmatrix}. \tag{33}$$

Bild 2 zeigt die drei Normalparabeln nach (18), (19) mit den Scheitelpunkten C_i:

$C_1\,(-0{,}4875/-0{,}2377)$,

$C_2\,(-1{,}9667/-2{,}8671)$,

$C_3\,(-0{,}4958/3{,}7541)$,

die Betragsgeraden mit den Steigungen $\pm R_i^2$:

$R_1^2 = 0{,}0553$; $R_2^2 = 0{,}0644$; $R_3^2 = 0{,}0380$

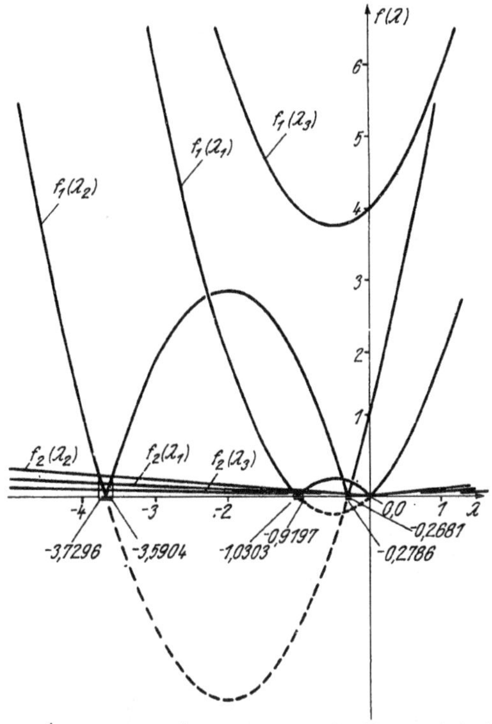

Bild 2. Einschließungsbereiche für die reellen Eigenwerte des Beispiels:

$-1{,}0303 \leqq \lambda_1 \leqq -0{,}9197$
$0{,}0 \leqq \lambda_2 \leqq 0{,}0$
$-3{,}7296 \leqq \lambda_3 \leqq -3{,}5904$
$-0{,}2786 \leqq \lambda_4 \leqq -0{,}2681$

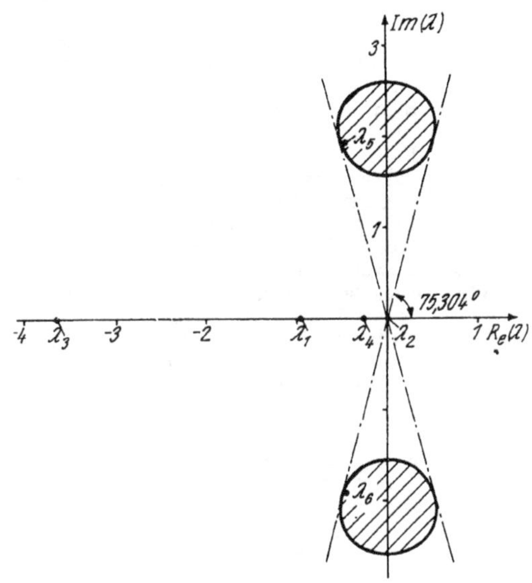

Bild 3. Exakte und Einschließungsbereiche für die komplexen Eigenwerte des Beispiels

und auf der reellen Achse die hieraus resultierenden Einschließungsbereiche. Da die dritte Parabel die zugehörige Gerade nicht schneidet, muß hier ein komplexer Einschließungsbereich existieren. Mit $S_3^2 = 1{,}0297$ sind wegen (31) getrennte Gebiete zu erwarten, die nach (29) zwischen

$$75{,}304° \,(+180°) \leqq \varphi \leqq 104{,}696° \,(+180°)$$

liegen, mit (27) berechnet und in Bild 3 eingezeichnet wurden wo auch die exakten Eigenwerte

$\lambda_1 = -0{,}97419$; $\lambda_2 = 0$;
$\lambda_3 = -3{,}66095$; $\lambda_4 = -0{,}27340$;
$\lambda_{5,6} = -0{,}49573 \pm i \cdot 1{,}9375$

eingetragen sind.

Literatur

1 FALK, S., Bemerkungen über parameternormale Matrizen und Matrizentupel, Abhandlungen der Braunschweigischen wissenschaftlichen Gesellschaft, im Druck.
2 CAUGHEY, T. K., Classical Normal Modes in Damped Linear Systems, J. Appl. Mech. 27, 269–271 (1960).

Eingereicht am 2. 5. 1979

Anschrift: Dipl.-Ing. JÖRG SCHNEIDER, Lehrstuhl für Mechanik und Festigkeitslehre im Mechanikzentrum der TU Braunschweig, Pockelsstraße 4, D-3300 Braunschweig, BRD

ZAMM 60, 268–270 (1980)

K. MLADENOV

Über das dynamische Verhalten eines Druckstabes mit tangententreuer Endbelastung

Einführung

Die Probleme des dynamischen Verhaltens elastischer Systeme bei polygen angreifenden Kräften gehören zur nichtkonservativen Mechanik, wobei die bisherigen Untersuchungen meist den geraden homogenen Druckstab bei verschiedenartigen Randbedingungen betreffen. Eine reiche Bibliographie zu dieser Frage sowie auch eine Reihe theoretischer Grundlagen kann man in [1], [2], [3], [4] finden. Zusammenfassungen einer Reihe experimenteller Resultate sowie die entsprechende Literatur werden in [5] gegeben.

Ein klassisches Beispiel für nichtkonservative Systeme ist das sogenannte BECKsche Problem [6], das in Abhängigkeit von den Randbedingungen statische (Divergenz) oder dynamische Instabilität (Flattern) zeigt. Bekannt ist, daß ein einseitig eingespannter Stab durch Flattern seine Stabilität verliert, wenn an seinem freien Ende eine Folgelast angreift, während ein beidseitig gelenkig gelagerter Stab bei einer an einem seiner Enden angreifenden polygenen oder monogenen Kraft ein- und denselben kritischen Wert für sie besitzt.

In der vorliegenden Arbeit wird das dynamische Verhalten einer Kombination dieser zwei Modelle untersucht, nämlich ein ursprünglich gerader Druckstab, gelenkig gelagert an einem Ende und in einem beliebigen Punkt, sowie belastet mit einer Kraft am freien Ende. Wir betrachten dazu die zwei in Bild 1a und Bild 1b gezeigten Anordnungen, im Folgenden entsprechend als „Fall a" und „Fall b" zitiert.

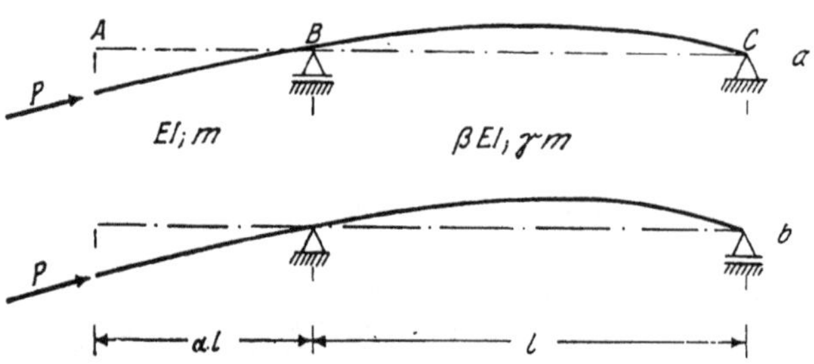

Bild 1. Knickstab mit Folgelast

Problemstellung

Die Differentialgleichung der elastischen Linie läßt sich mittels des Prinzips von D'ALEMBERT-LAGRANGE ableiten. Diese Prozedur wird sehr häufig in der zitierten Literatur angewendet, weswegen hier nur das Endergebnis gegeben wird. Wenn man mit $w(x, \tau)$ die Verschiebung der Punkte der Stabachse bezeichnet, mit x die Koordinate in Achsenrichtung, mit τ die Zeit, und sich auf kleine Verschiebungen und Winkel beschränkt, so erhält man folgende Differentialgleichungen für die Bewegung des Stabs bei Vernachlässigung der Rotationsenergie:

$$E_1 I_1 w_1''''(x_1, \tau) + P w_1''(x_1, \tau) + m_1 \ddot{w}_1(x_1, \tau) = 0 ,$$
$$E_2 I_2 w_2''''(x_2, \tau) + P w_2''(x_2, \tau) + m_2 \ddot{w}_2(x_2, \tau) = 0 \tag{1}$$

Dabei sind m_1 und m_2 spezifische Massen des Stabes, $E_1 I_1$ und $E_2 I_2$ als konstant angenommene Biegesteifigkeiten entsprechend den Teilstücken AB und BC. Dieses System von Differentialgleichungen wird bei folgenden Randbedingungen gelöst:

$$x_1 = 0: \quad w_1''(0, \tau) = w_1'''(0, \tau) = 0;$$
$$x_1 = \alpha l, (x_2 = 0): \quad w_2(0, \tau) = 0 , \qquad w_1'(\alpha l, \tau) = w_2'(0, \tau) ,$$
$$w_1(\alpha l, \tau) = 0 , \qquad E_1 I_1 w_1''(\alpha l, \tau) = E_2 I_2 w_2''(0, \tau);$$
$$x_2 = l: \quad w_2(l, \tau) = w_2''(l, \tau) = 0 . \tag{2}$$

Zur Durchführung der Zahlenrechnung sei

$$E_1 I_1 = EI , \qquad E_2 I_2 = \beta EI , \qquad m_1 = m , \qquad m_2 = \gamma m$$
$$0 \leq x_1 \leq \alpha l , \qquad 0 \leq x_2 \leq l \tag{3}$$

gewählt, wobei α, β und γ gegebene Koeffizienten sind. Mit den dimensionslosen Größen

$$\xi_i = x_i/l , \qquad \eta_1 = \eta = Pl^2/(EI) , \qquad \eta_2 \beta = \eta$$
$$y_i = w_i/l , \qquad \tau^2 = t^2 ml^4/(EI) , \qquad i = 1, 2, \tag{4}$$

wird man auf die zugehörige Eigenwertaufgabe

$$y_1''''(\xi_1, t) + \eta y_1''(\xi_1, t) + \ddot{y}_1(\xi_1, t) = 0 , \qquad 0 \leq \xi_1 \leq \alpha$$
$$y_2''''(\xi_2, t) + \eta_2 y_2''(\xi_2, t) + \frac{\gamma}{\beta} \ddot{y}_2(\xi_2, t) = 0 , \qquad 0 \leq \xi_2 \leq 1 , \tag{5}$$

$$\xi_1 = 0: \quad y_1''(0, t) = y_1'''(0, t) = 0;$$
$$\xi_1 = \alpha, (\xi_2 = 0): \quad y_1(\alpha, t) = y_2(0, t) = 0 , \tag{6}$$
$$y_1'(\alpha, t) = y_2'(0, t) , \qquad y_1''(\alpha, t) = \beta y_2''(0, t) ,$$
$$\xi_2 = 1: \quad y_2''(1, t) = y_2(1, t) = 0$$

geführt. Hierbei bedeutet $(\ )'$ Differentiation nach ξ_j, und $(\dot{\ })$ Differentiation nach der dimensionslosen Zeit t.

Mittels des Produktansatzes

$$y_j(\xi_j, t) = X_j(\xi_j) \, e^{i\omega t} , \qquad j = 1, 2 , \tag{7}$$

wobei ω ein dimensionsloser Frequenzparameter ist, verbunden mit der wirklichen Frequenz Ω durch die Beziehung

$$\omega^2 = \Omega^2 ml^4/(EI) , \tag{8}$$

bekommt man die gewöhnlichen Differentialgleichungen

$$X_j^{IV}(\xi_j) + \eta_j X_j''(\xi_j) - \varepsilon_j^4 X_j(\xi_j) = 0 , \qquad j = 1, 2 . \tag{9}$$

Mit den Abkürzungen

$$\delta_j, \gamma_j = \sqrt{z_j \pm \eta_j/2} , \quad z_j = \sqrt{\eta_j^2/4 + \varepsilon_j^4} , \quad \varepsilon_1^4 = \varepsilon^4 = \omega^2, \quad \varepsilon_2^4 = \varepsilon^4 \gamma/\beta \tag{10}$$

lauten die allgemeinen Lösungen

$$X_j(\xi_j) = C_{j1} \operatorname{ch} \gamma_j \xi_j + C_{j2} \operatorname{sh} \gamma_j \xi_j + C_{j3} \cos \delta_j \xi_j + C_{j4} \sin \delta_j \xi_j . \tag{11}$$

Die Randbedingungen (6) geben acht homogene lineare Gleichungen zur Ermittlung der acht Konstanten C_{jk} ($j = 1, 2$; $k = 1, 2, 3, 4$). Als Bedingung für die Existenz nichttrivialer Lösungen erhält man die Eigenwertgleichung

$$\varepsilon^2 z_1 (\delta_1 \operatorname{sh} \gamma_1 \alpha \cos \delta_1 \alpha - \gamma_1 \operatorname{ch} \gamma_1 \alpha \sin \delta_1 \alpha) (\delta_2 \operatorname{sh} \gamma_2 \cos \delta_2 -$$
$$- \gamma_2 \operatorname{ch} \gamma_2 \sin \delta_2) - z_2 \beta [\gamma_1^4 + \delta_1^4 + \varepsilon^2 (\eta \operatorname{sh} \gamma_1 x \sin \delta_1 \alpha +$$
$$+ 2\varepsilon^2 \operatorname{ch} \gamma_1 \alpha \cos \delta_1 \alpha] \operatorname{sh} \gamma_2 \sin \delta_2 = 0 . \tag{12}$$

Interessant ist, daß für $\alpha = 0$ die obige Gleichung die Form

$$\operatorname{sh} \gamma_2 \sin \delta_2 = 0 \tag{13}$$

annimmt, welche für $\omega = 0$ den EULERschen kritischen Wert für einen beidseitig gelenkig gelagerten Stab ergibt, und für $P = 0$ — die Frequenzen der Eigenschwingungen für den gleichen Fall.

Die charakteristische Gleichung (12), sowie auch alle bisherigen Untersuchungen beziehen sich auf den „Fall a". Bei „Fall b" hat die Differentialgleichung (9) für das Teilstück BC die Form

$$X_2''''(\xi_2) - \varepsilon_2^4 X_2(\xi_2) = 0 \tag{14}$$

mit der Lösung

$$X_2(\xi_2) = C_{21}\,\mathrm{ch}\,\varepsilon_2\xi_2 + C_{22}\,\mathrm{sh}\,\varepsilon_2\xi_2 + C_{23}\cos\varepsilon_2\xi_2 + C_{24}\sin\varepsilon_2\xi_2\,. \tag{15}$$

Die entsprechende charakteristische Gleichung ist

$$\varepsilon^2 z_1(\delta_1\,\mathrm{sh}\,\gamma_1\alpha\cos\delta_1\alpha - \gamma_1\,\mathrm{ch}\,\gamma_1\alpha\sin\delta_1\alpha)\,(\mathrm{sh}\,\varepsilon_2\cos\varepsilon_2 - \mathrm{ch}\,\varepsilon_2\sin\varepsilon_2) - \varepsilon_2\beta[\gamma_1^4 + \beta_1^4 + \varepsilon^2(\eta\,\mathrm{sh}\,\gamma_1\alpha\sin\delta_1\alpha + 2\varepsilon_1^2\,\mathrm{ch}\,\gamma_1\alpha\cos\delta_1\alpha)] \times \times\,\mathrm{sh}\,\varepsilon_2\sin\varepsilon_2 = 0\,. \tag{16}$$

Numerische Ergebnisse. Diskussion

Die Lösung der transzendenten Gleichungen (12) und (16) wurde mit einer elektronischen Rechenmaschine durchgeführt, wobei die Parameter α, β und γ variiert wurden. Die Resultate der Lösung werden in den Bildern 2 bis 8 dargestellt. Im Bild 2 wird die Abhängigkeit zwischen dem Belastungsparameter η und dem der Frequenz ε für verschiedene Werte des Koeffizienten α und $\beta = \gamma = 1$ gezeigt. Das sind die sogenannten Eigenwertkurven. Es erweist sich, daß bei „Fall a" dem statischen wie auch dem dynamischen Stabilitätsverlust entsprechende kritische Werte für den Parameter η existieren. Dabei entspricht — in Abhängigkeit von α — der minimale kritische Wert von η sowohl dem Flattern als auch der Divergenz. Das wird in Bild 4 gezeigt, bei dem die drei kleinsten kritischen Werte von η in Abhängigkeit von α für $\beta = \gamma = 1$ aufgetragen sind. Zu bemerken ist, daß für α in den Grenzen von Null bis annähernd 1,37 und $\beta = \gamma = 1$ der Ausdruck min η_{kr} konstant gleich π^2 und nicht von α abhängig ist. Für $\alpha > 1{,}37$ verringert sich min η_{kr} mit dem Anwachsen von α.

Bild 3 zeigt die Eigenwertkurven für $\beta = \gamma = 1$ und „Fall b". Es erweist sich, daß hier lediglich dynamische Instabilität (Flattern) herrscht. Die entsprechende graphische Darstellung im Bild 4 zeigt, daß für $\alpha > 1{,}37$ der minimale kritische Parameter der Belastung für die Fälle a und b praktisch derselbe ist.

Der Einfluß des Parameters β auf die Dynamik des Druckstabes im „Fall a" für $\alpha = \gamma = 1$ wird in den Bildern 5 und 7 gezeigt. Mit dem Anwachsen von β, d. h. mit dem Steigen der Biegesteifigkeit im Abschnitt des Stabes zwischen den Stützen im Verhältnis zum Abschnitt AB, wächst auch min η_{kr} entsprechend der Divergenz. Die entsprechenden Werte sind

$$\eta_{kr} = \beta\pi^2\,. \tag{17}$$

Zu bemerken ist, daß für $\beta < 0{,}42$ sowohl der erste als auch der zweite kritische Wert von η der statischen Instabilität entspricht, während für $\beta > 0{,}42$ der zweite kritische Wert dem Flattern entspricht. Für $\beta > 2{,}02$ entspricht min η_{kr} dem Flattern. Mit dem Anwachsen von β über diese Grenze hinaus wächst min η_{kr} ebenfalls und bei $\beta \to \infty$ erreicht η_{kr} den Wert von 20,05, bekannt als Becksche Lösung [6]. Man kann sagen, daß man für $\beta > 10{,}0$ praktisch das Becksche Modell anstelle des hier betrachteten komplizierteren Beispiels annehmen kann. So beträgt z. B. der Fehler für η_{kr} für $\beta = 10{,}5$ ungefähr 6%, während $\beta = 5{,}8$ zu einem Fehler von ca. 14% führen würde.

In Bild 6 sind die Eigenwertkurven für $\alpha = \beta = 1$ und variierendem Koeffizienten γ für „Fall a" dargestellt. Für jedes γ des Intervalls $0{,}25 \leqq \gamma \leqq 100{,}0$ gilt min $\eta_{kr} = \pi^2$, was dem Fall der Divergenz entspricht. Für $\gamma = 0{,}25$; 2; 4 unterscheiden sich die ersten Eigenwertkurven geringfügig von der bei $\gamma = 1$ und sind deshalb nicht im Bild festgehalten. Darüber hinaus zeigt Bild 6 den schwachen Einfluß von γ auf den kritischen Parameter η, was dem Fall des Flatterns entspricht, das im betrachteten Fall ($\alpha = \beta = 1$) nicht minimal ist. So z. B. ist $\eta_{kr} = 19{,}12$ für $\gamma = 0{,}25$ und $\eta_{kr} = 17{,}05$ für $\gamma = 100{,}0$.

Die ersten vier Eigenformen im „Fall a" für $\alpha = 1{,}40$, $\beta = \gamma = 1$, und bei den Werten $\eta = 4{,}0$; 8,5 und 9,5, d. h. bei relativ kleinem η, bei η nahe dem ersten kritischen Wert und bei η nahe dem zweiten kritischen Wert, sind im Bild 8 a, b, c dargestellt. Dabei wird angenommen, daß für jede Kurve die Verschiebung des Punktes $\xi_1 = 0$ gerade 1 beträgt. Man sieht, daß sich die erste, dritte und vierte Eigenform qualitativ nicht verändern, während sich die zweite Eigenform der ersten annähert bei Annäherung von η an den dem Flattern entsprechenden kritischen Wert. Demzufolge entsteht die Frage, ob sich der Druckstab nicht in der Zone der dynamischen Instabilität befindet, noch bevor η den erwähnten kritischen Wert erreicht hat. Auf Grund dessen, daß die zwei benachbarten Eigenformen der Art und den entsprechenden absoluten Größen der Eigenwerte nach ähnlich sind, kann man annehmen [7], daß dynamische Instabilität in Form der Schwebung vor der Erscheinung des Flatterns auftritt. Die Bestätigung dieser Behauptung oder deren Widerlegung erfordert zusätzliche Untersuchungen, nicht nur für das hier betrachtete Modell, sondern auch für das Becksche Modell.

Literatur

1 Болотин, В. В., Неконсервативные задачи теории упругой устойчивости, Наука, Москва, 1961.
2 Leipholz, H., Stabilitätstheorie, B. G. Teubner 1968.
3 Leipholz, H., Direct variational methods and eigenvalue problems in engineering, Noordhoff int. publ., Leyden, 1977.
4 Herrmann, G., Stability of equilibrium of elastic systems subjected to nonconservative forces, AMR **20**, No. 2, 1967.
5 Sugiyama, Y., Sekiya, Ts., Surveys of the experimental studies on instability of the elastic systems subjected to nonconservative forces (in Japanisch), Journal of the Japan Society for Aeronautical and Space Sciences **19**, 61–68 (1971).
6 Beck, M., Die Knicklast des einseitig eingespannten, tangential gedrückten Stabes, ZAMP **3**, 225–228 (1962).
7 Den Hartog, J. P., Vibrations mécaniques, Dunod 1960.

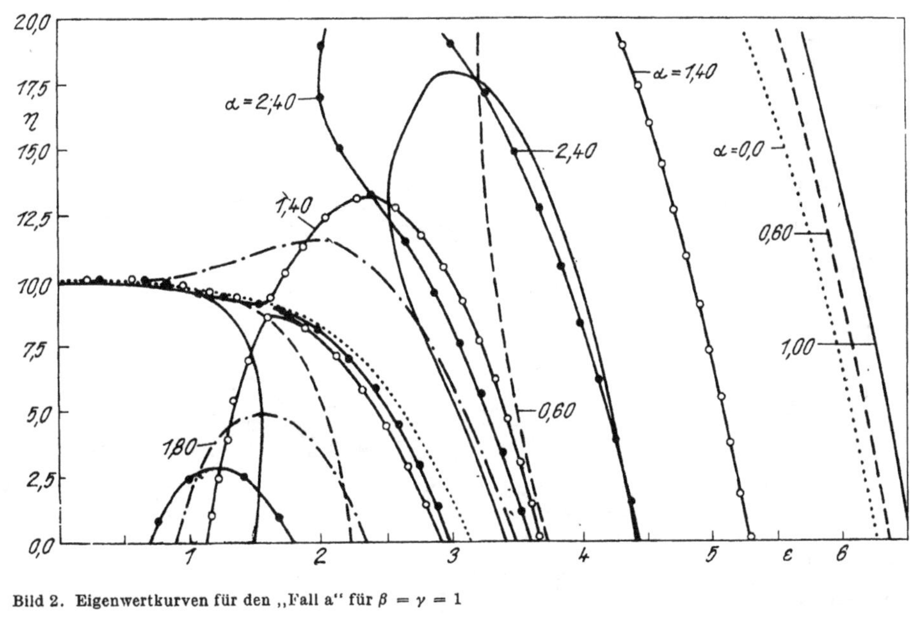

Bild 2. Eigenwertkurven für den „Fall a" für $\beta = \gamma = 1$

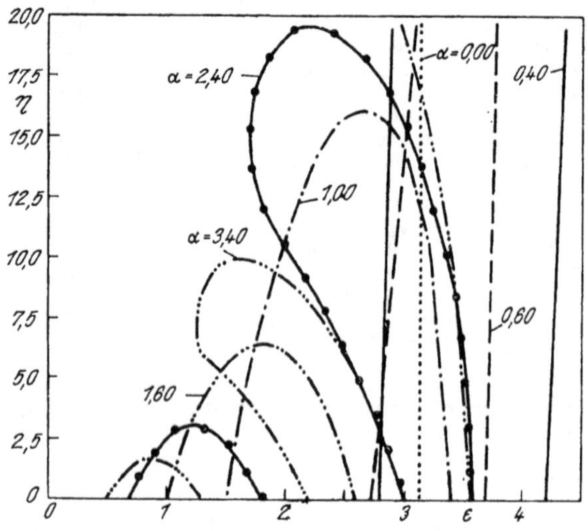

Bild 3. Eigenwertkurven für den „Fall b" für $\beta = \gamma = 1$

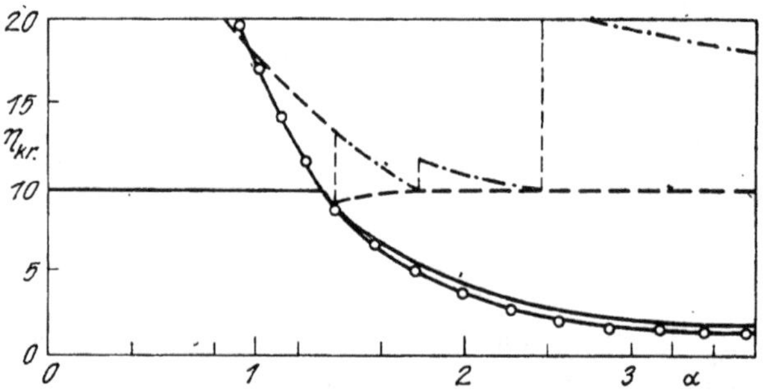

Bild 4. Stabilitätskarte. Für den „Fall a": ——— erster kritischer Wert von η — — — zweiter kritischer Wert, —·—·— dritter kritischer Wert. Für den „Fall b": —○—○— erster kritischer Wert, $\beta = \gamma = 1$

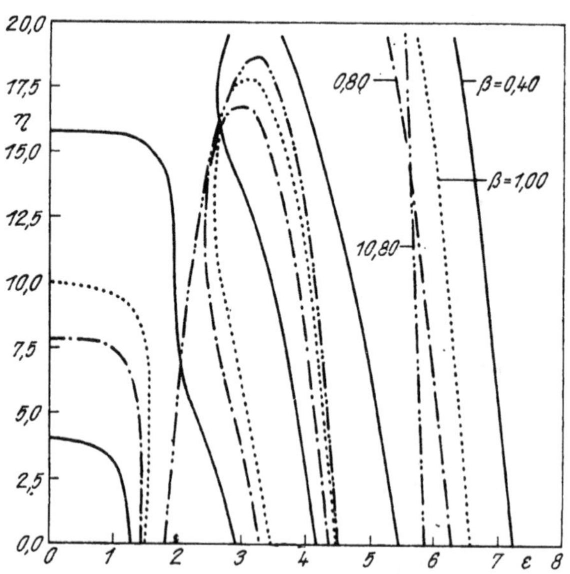

Bild 5. Eigenwertkurven für den „Fall a" für $\alpha = \gamma = 1$

Eingereicht am 29. 11. 1978

Anschrift: Dr.-Ing. K. Mladenov, TH für Bauwesen und Architektur, Lehrstuhl Technische Mechanik, Hr. Smirnensky Str. 1, Sofia 1421, Bulgarien

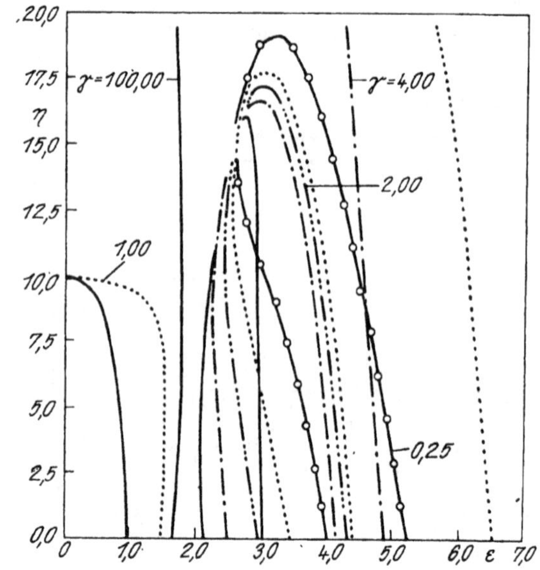

Bild 6. Eigenwertkurven für den „Fall a" für $\alpha = \beta = 1$

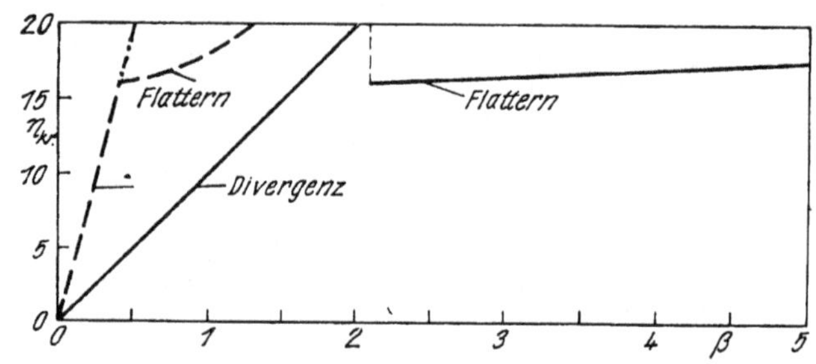

Bild 7. Stabilitätskarte für den „Fall a" für $\alpha = \gamma = 1$: ——— erster kritischer Wert, — — — zweiter kritischer Wert, —·—·— dritter kritischer Wert

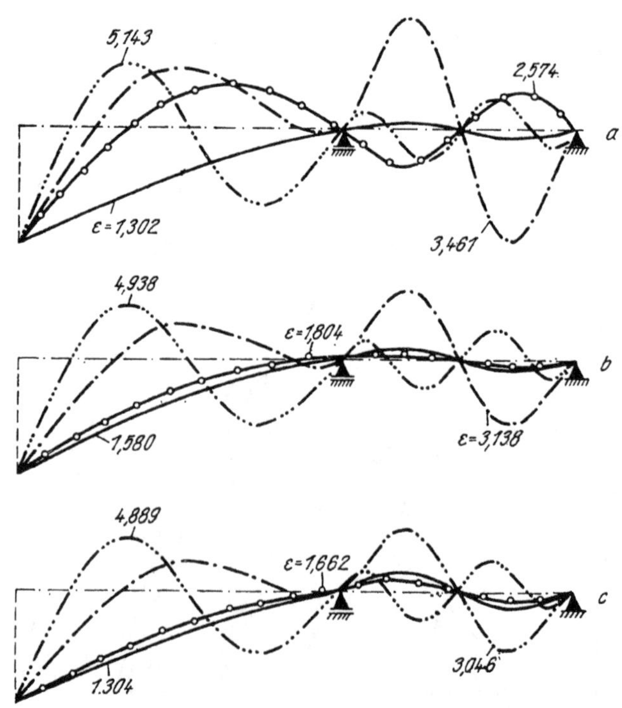

Bild 8. Eigenformen für den „Fall a" für $\alpha = 1{,}40$, $\beta = \gamma = 1$; a) $\eta = 4{,}0$ b) $\eta = 8{,}5$; c) $\eta = 9{,}5$. ——— 1. Eigenform, —○—○— 2. Eigenform, —·—·— 3. Eigenform, ···—···— 4. Eigenform

BUCHBESPRECHUNGEN

Klein, F., Vorlesungen über die Entwicklung der Mathematik im 19. Jahrhundert. Ausgabe in einem Band. Reprint der Erstauflage von Teil 1 und 2, Berlin 1926 und 1927. Berlin-Heidelberg-New York, Springer-Verlag 1979. XV, 385 und IX, 208 S., 55 Abb., DM 36,00, US $ 19,80. ISBN 3-540-09235-8

Man hat dem Verlag zu danken für die Initiative zum Nachdruck. F. KLEIN hat mit diesem Buch einen sehr nachhaltigen Einfluß auf die Historiographie der Mathematik während des 20. Jahrhunderts ausgeübt. Der Reiz des Buches bestand und besteht — neben der Vermittlung einer Fülle von Details und historischen Tatsachen — gerade eben in dem Umstande, daß ein erfolgreicher, produktiver Mathematiker und hervorragender Wissenschaftsorganisator über historische Entwicklungen urteilt, die er größtenteils als Zeitgenosse erlebt und sogar selbst mit seiner eigenen Leistung gestaltet hat. — KLEINS „Vorlesungen ...“ standen damals an der Berührungsfläche von Historiographie und retenter Forschung; heute ist erwiesen, daß dieses sein Werk zur grundlegenden mathematik-historischen Literatur zu rechnen ist. Das schließt nicht aus, daß wir heute einige Aussagen von KLEIN als subjektiv, als zeitbedingt, als durch den Geschichtsverlauf korrigiert einzuschätzen haben, etwa eine bei ihm spürbare Art von Geringschätzung der abstrakten, axiomatischen Methode (S. 336) und der Mengentheorie. Wichtiger aber ist der durchweg zutage tretende, faszinierende Sachverstand und die ungemein lebendige Form der Darstellung. — Und es muß und kann hervorgehoben werden, daß KLEIN, selbst ein engagierter Verfechter der angewandten Mathematik, gerade auch der Geschichte der Anwendungen der Mathematik in Mechanik, Optik, Elektrodynamik, Relativitätstheorie und Technik breiten Raum einräumt; so sind und bleiben KLEINS „Vorlesungen ...“ eine reiche Fundgrube auch für den historisch interessierten angewandten Mathematiker.

Leipzig H. WUSSING

Jeggle, H., Nichtlineare Funktionalanalysis, Existenz von Lösungen nichtlinearer Gleichungen. Teubner Studienbücher Mathematik. Stuttgart, B. G. Teubner 1979. 255 S., 52 Aufg., DM 24,80. ISBN 3-519-02057-2

Das Buch umfaßt die folgenden Kapitel: 1. Vorbereitendes, 2. Kontraktive und nichtexpansive Operatoren, 3. Abbildungsgrade und kompakte Operatoren, 4. Variationsmethoden, 5. Monotone Operatoren, 6. Fixpunktsätze und mengenwertige Operatoren, dazu ein Literaturverzeichnis mit 162 Titeln, die meisten aus den sechziger und siebziger Jahren. Der Leser sollte über Grundkenntnisse in der Funktionalanalysis verfügen und in der Lage sein, knapp dargestellte Sachverhalte zu verstehen; dann hilft ihm der vorliegende Text, tiefer in die Behandlung nichtlinearer Probleme einzudringen. Besonders hilfreich sind dabei die Motivationen für abstrakte Begriffsbildungen. Durch die jedem Kapitel angefügten Übungsaufgaben, Bemerkungen und Hinweise wird der Leser zu selbständiger Mitarbeit angeregt, und es werden Verbindungen zur Spezialliteratur und zum gegenwärtigen Stand der Forschung hergestellt. Das hier dargestellte Gebiet ist für die Anwendungen der Mathematik in Mechanik und Physik von aktueller Bedeutung, das bezieht sich besonders auf die Grundgleichungen der Kontinuumsmechanik und der Plasmaphysik. Man möchte wünschen, daß das Buch auch in diesen Fachkreisen Interesse findet; Anknüpfungspunkte sind vorhanden. Z. B. findet man allein das Stichwort Randwertaufgaben an 5 Stellen. Allerdings sagt der Verfasser selbst, daß das Konzept dieses Buches aus Vorlesungen und Seminaren entstanden ist, die er im Rahmen eines Zyklus über Funktionalanalysis und Numerische Mathematik im Mathematik-Hauptstudium an der TU in Charlottenburg gehalten hat.

Berlin J. FÖRSTE

Geel, R., Singular Perturbations of Hyperbolic Type. Mathematical Centre Tracts 98. Amsterdam, Mathematisch Centrum 1978. 184 S., Dfl. 23.—. ISBN 90-6196-166-1

Sind L_1, L_2 Differentialoperatoren und ist ε ein kleiner Parameter, so heißt ein Differentialgleichungsproblem für $L_1 + \varepsilon L_2$ (anstelle für L_1) bekanntlich gestört. Ist die Ordnung von L_2 nicht größer als die von L_1, so heißt die Störung regulär, andernfalls singulär. Der schwierige Fall ist natürlich der der singulären Störungen, weil das sachgemäße Problem für $L_1 + \varepsilon L_2$ anders als für den ungestörten Operator L_1 zu stellen ist. In nahezu axiomatischer Form zeigt der Verfasser in einer ausführlichen Einführung, wie (bei kleinen ε) eine Fehlerabschätzung für formale Approximationen der Lösung des gestörten Problems auf a-priori-Abschätzungen zurückgeführt werden kann. Letztere werden durch Energieintegral-Methoden gewonnen.

Der eigentliche Gegenstand der Monographie sind Differentialgleichungen in der Ebene, wobei L_1 von erster Ordnung und L_2 von zweiter und überdies hyperbolisch ist. Da alle charakteristischen Erscheinungen für die betrachteten singulär gestörten Probleme bereits bei gewöhnlichen Differentialgleichungen auftreten, wird zunächst das singulär gestörte Anfangswertproblem für gewöhnliche Differentialgleichungen (L_1 von erster, L_2 von zweiter Ordnung) behandelt. Kapitel I ist linearen, Kapitel II quasilinearen Differentialgleichungen gewidmet.

Nach diesen vorbereitenden Kapiteln über gewöhnliche Differentialgleichungen bringt Kapitel III das singulär gestörte Anfangswertproblem für lineare Differentialoperatoren. Von der Lösung, die in der oberen Halbebene gesucht wird, werden die Werte und die der ersten Ableitungen auf dem Rand vorgeschrieben. Während das Existenzgebiet der Lösung spezielle Gestalt haben soll, können die Operatoren beliebige lineare sein (L_2 hyperbolisch). Im folgenden Kapitel IV wird zugelassen, daß die Koeffizienten von L_1 von der gesuchten Lösung selbst abhängen dürfen. Kapitel V setzt wieder Linearität der Differentialgleichung voraus, das Existenzgebiet ist hier aber im Gegensatz zu Kapitel III ein streifenförmiges Gebiet. Für die gesuchte Lösung wird ein Rand-Anfangswertproblem gestellt (u und seine erste Ableitung ist für $t = 0$ vorgesehen, während für alle t die Randwerte von u vorgegeben sind).

Während $L_2[u]$ in den Kapiteln IV und V als $\partial^2 u/\partial t^2 - c^2(x, t) \cdot \partial^2 u/\partial x^2$ vorausgesetzt ist, wird $L_2[u]$ im letzten Kapitel VI als $\partial^2 u/\partial x \partial t$ angenommen. Dann ist die x-Achse eine Charakteristik. Für ein von der x-Achse und von einer durch $x = \Phi(t)$ definierten Kurve (Φ_2 nicht fallend, $\Phi_2(0) = 0$) berandetes Gebiet wird in Kapitel VI dann für u das charakteristische Randwertproblem (GOURSAT-Problem) gestellt.

Das Buch bringt nicht nur eigene Resultate des Verfassers, es kann gleichzeitig als Einführung in die Technik der asymptotischen Darstellungen bei singulären Störungen angesehen werden. Als seine besonderen Vorzüge seien genannt:

a) alle für die Herleitung der erhaltenen Ergebnisse erforderlichen Voraussetzungen werden übersichtlich und vollständig angegeben,

b) die Ergebnisse werden in klar formulierten Sätzen zusammengefaßt, die auch alle erforderlichen Voraussetzungen nennen,

c) für Spezialfälle werden die erhaltenen Ergebnisse noch weiter konkretisiert,

d) Hinweise auf Anwendung der Theorie bei Verkehrsströmen; Ströme in langen Flüssen usw.) werden gegeben.

Halle W. TUTSCHKE

Page, E. S. / Wilson, L. B., Information Representation and Manipulation in a Computer. Second Edition. Cambridge Computer Science Texts 2. Cambridge-London-New York-Melbourne, Cambridge University Press 1979. 271 S., £ 10.00 A H/C, £ 3.95 A P/B. ISBN 0521220882 H/C; 0521293357 P/B

Die Autoren halten es bei der raschen Entwicklung der Anwendungen der Informatik für erforderlich, daß die damit Befaßten nicht nur Hilfsmittel wie etwa eine höhere Programmierungssprache für die Lösung praktischer Probleme kennenlernen, sondern auch einiges über die theoretischen Grundlagen wissen. Dazu soll das Buch eine Einführung bieten. Es setzt demgemäß an Vorkenntnissen nur elementare Mathematik und etwas Kenntnis der Programmierung voraus. Behandelt werden die wichtigsten Darstellungen von Daten und einigen rechentechnischen Grundstrukturen wie Felder, Listen und Bäume und Grundaufgaben wie Suchen und Sortieren. Dabei werden typische Anwendungsfälle genannt und Vergleiche (mit Hinweis auf deren notwendigerweise unvollständigen Charakter) gezogen.

Jedes Kapitel enthält ein ausführliches Verzeichnis weiterführender Literatur und eine Fülle von Aufgaben sehr verschiedenen Schwierigkeitsgrades (vielfach Universitätsprüfungen entnommen), zu deren Lösung im Anhang Hinweise gegeben sind. Das Buch ist gut lesbar, für den ins Auge gefaßten Interessentenkreis erscheint allerdings das Kapitel 1 (Art und Bedeutung von Symbolen) überflüssig, weil bekannt.

Berlin W. ISSEL

Keilson, J., Markov Chain Models — Rarity and Exponentiality. Applied Mathematical Sciences 28. Berlin-Heidelberg-New York, Springer-Verlag 1979. XIII, 184 S., DM 24,—, US $ 13,20. ISBN 3-540-90405-0

The author briefly introduces basic tools like ergodic state probabilities, ergodic flow rates, ruin probabilities, passage time and regeneration time distributions. At length he considers time reversible chains and the exploitation of spatial homogeneity and independence underlying certain "modified processes" Finally emphasis is given to rarity and exponentiality, i.e., to the limit theorems concerning these, to measures of exponentiality and error bounds associated with such measures. Potential theory and the author's compensation method are also treated. Remarkable is also a 10-page-treatment of stochastic monotonicity. Applications are given mainly to reliability theory and inventory theory. Unfortunately motivations and intuitive explanations are few.

Leipzig H.-J. ROSSBERG

Schwefel, H.-P., Numerische Optimierung von Computer-Modellen mittels der Evolutionsstrategie. Mit einer vergleichenden Einführung in die Hill-Climbing- und Zufallsstrategien. (ISR 26) Basel-Stuttgart, Birkhäuser Verlag 1977. 390 S., sFr. 48,—.

Vorgelegt wird hier eine umfassende Beschreibung und Bewertung moderner numerischer Methoden der Optimierung von Computer-Modellen. Die Parameteradaptation in algorithmischen Simulationsmodellen führt i. a. auf Optimierungsprobleme mit unbekannter Struktur des Zielfunktionals. Für solche Probleme behandelt der Verfasser geeignete numerische Lösungsverfahren: direkte Suchverfahren, einschließlich einer besonderen Art von stochastischen Verfahren, den Evolutionsstrategien. (Evolutionsstrategien sind direkte, stochastische Suchverfahren, die nach Prinzipien der biologischen Evolution aufgebaut sind.) Herausgehoben werden sollen die folgenden inhaltlichen Schwerpunkte: 1. Die eingehende Beschreibung von direkten Suchverfahren unter dem Blickwinkel ihrer rechentechnischen Brauchbarkeit. Dazu zählen Methoden der eindimensionalen Minimierung und Methoden wie: Koordinatenstrategien, Direkte Suche von HOOKE und JEEVES, DAVIES-SWANN-CAMPEY-Methode, POWELLS Verfahren der konjugierten Richtungen, DAVIDON-FLETCHER-POWELL-Version der Methode der variablen Metrik, Complex-Verfahren von Box. 2. Grundlegende Ausführungen zur Konzeption und numerischen Effektivität von Evolutionsstrategien. Aufgebaut wird auf die richtungsweisenden Überlegungen von RECHENBERG, die unter Beachtung numerisch-rechentechnischer Aspekte fortgeführt werden und in die Vorstellung einer eigenen „mehrgliedrigen Evolutionsstrategie", kurz: MGE-Strategie, münden. 3. Umfassende numerische Untersuchungen zu Effektivität, Zuverlässigkeit und globaler Konvergenzeigenschaft der vorgestellten Methoden an einem Spektrum von Testaufgaben. Herausgehoben wird in den vergleichenden Untersuchungen, daß die Evolutionsstrategien in ihrer Effektivität durchaus mit den deterministischen Verfahren konkurrieren können und ihnen hinsichtlich Zuverlässigkeit und (natürlich) globalem Konvergenzverhalten überlegen sind.

Vom Erscheinen an ab hat das Buch ein lebhaftes Interesse gefunden — einschließlich kritischer Wertung der EGL-Methode. Es ist eine wertvolle Arbeitsgrundlage für alle, die mit der Auswahl oder Entwicklung numerisch effektiver Optimierungsverfahren beschäftigt sind.

Berlin J. BORN

Böhling, K. H. / Spies, P. P. (Hrsg.), GI — 9. Jahrestagung. Bonn, 1.—5. Oktober 1979. Informatik-Fachberichte 19. Berlin-Heidelberg-New York, Springer-Verlag 1979. XII, 690 S., 133 Abb., 25 Tab., DM 56,60, US $ 31.10. ISBN 3-540-09664-7

Die 9. Jahrestagung der Gesellschaft für Informatik stellte sich das Ziel, einen Überblick über den gegenwärtigen Stand der Grundlagen und Anwendungen der Informatik zu vermitteln.

Der vorliegende Band bringt 57 Beiträge. Ein zusätzlicher Bericht für die Tagungsteilnehmer enthält insbesondere die Kurzvorträge. In Hauptvorträgen werden Geschichte, Gegenwart und Zukunft der Informatik (HÄNDLER), Konzepte der Komplexitätstheorie (MEHLHORN), Probleme der Systemprogrammierung (SCHNUPP), die rechnergestützte Automobilentwicklung (LINCKE u. SIEPMANN), die Entwicklung der Programmiersprache ADA (WICHMANN), Datenbank-Anfragesprachen (LEHMANN u. BLASER) und Anwendungen der Netztheorie (PETRI) abgehandelt. Zwei Beiträge befassen sich mit Möglichkeiten der Verrechtlichung der Datenverarbeitung und mit rechtlichen Fragen der internationalen Datenkommunikation. Im Rahmen der Sitzung „Spezifikation von Echtzeitsystemen" werden in 6 Arbeiten Fragen des Entwurfs und der Herstellung von Prozeß-Software sowie der Modellierung angeschnitten. Zum Fachgespräch „Compiler — Compiler" finden sich 8 Beiträge, die zum einen allgemeine theoretische und zum anderen system-spezifische Aspekte behandeln. In 7 Vorträgen des Fachgesprächs „Methoden- und Modellbankensysteme" werden Architekturprinzipien, Zugriffsformen, Einsatzmöglichkeiten und spezielle Systeme beleuchtet. Das Fachgespräch „Modelle für Rechnersysteme" diskutiert in 13 Referaten Probleme der Analyse des Leistungsverhaltens und der Modellierung von Systemen. In 14 Arbeiten des Fachgesprächs „Organisation und Betrieb von Rechenzentren" werden praktische Fragen erörtert.

Berlin H. W. MEIER

EINGEGANGENE BÜCHER

Bei der Schriftleitung sind folgende Bücher eingegangen (ausführliche Besprechung bleibt vorbehalten).

Jain, M. K., Numerical Solution of Differential Equations. New Delhi-Bangalore-Bombay-Calcutta, Wiley Eastern Limited 1979. XIII, 443 S., 45 Abb., 49 Tab., £ 7.50, US $ 17.65. ISBN 0-85226427-5

Igusa, J.-I., Lectures on Forms of Higher Degree. Notes by **S. Raghavan.** Tata Institute of Fundamental Research Lectures on Mathematics and Physics 59. Berlin-Heidelberg-New York, Springer-Verlag 1978. VI, 175 S., 1 Abb., DM 18,–, US $ 8.00. ISBN 3-540-08944-6

Barut, A. O. (Ed.), Nonlinear Equations in Physics and Mathematics. Proceedings of the NATO Advanced Study Institute held in Istanbul, Turkey, August 1–13, 1977. NATO Advanced Study Institute Series. Series C – Mathematical and Physical Sciences 40. Dordrecht-Boston-London, D. Reidel Publishing Company 1977. X, 473 S., 70 Abb., 2 Tab., Dfl. 80,– US $ 39.00. ISBN 90-277-0936-X

Fetzer, A. / Fränkel, H., Mathematik. Lehrbuch für Fachhochschulen, Band 3. Hannover-Dortmund-Darmstadt-Berlin, Hermann Schroedel Verlag KG 1979. 423 S., 260 Abb., 11 Tab., DM 44,80. ISBN 3-507-99047-4

Löhr, H. J., Beispiele und Aufgaben zur Laplace-Transformation mit 167 Beispielen und 37 Aufgaben. Viewegs Fachbücher der Technik. Braunschweig-Wiesbaden, Friedr. Vieweg & Sohn 1979. VI, 202 S., 61 Abb., 1 Tab., DM 24,80. ISBN 3-528-04082-3

Gordon, M. J. C., The Denotational Description of Programming Languages. An Introduction. Berlin-Heidelberg-New York, Springer-Verlag 1979. V, 160 S., DM 18,60, US $ 10.30. ISBN 3-540-90433-6

Lazarević, B. (ed.), Global and Large Scale System Models. Proceedings of the Center for Advanced Studies (CAS) – International Summer Seminar Dubrovnik, Yugoslavia, August 21–26, 1978. Sponsored by: The Faculty of Organizational Sciences, University of Belgrade, Yugoslavia. Lecture Notes in Control and Information Sciences 19. Berlin-Heidelberg-New York, Springer-Verlag 1979. VIII, 232 S., 36 Abb., 4 Tab., DM 25,–, US $ 13.80. ISBN 3-540-09637-X

Siegert, H. J. (Hrsg.), Virtuelle Maschinen. Nachbildung und Vervielfachung maschinenorientierter Schnittstellen. GI-Arbeitsseminar, München 1979. Informatik-Fachberichte 18. Berlin-Heidelberg-New York, Springer-Verlag, 1979. X, 230 S., 62 Abb., 5 Tab., DM 25,–, US $ 13.80. ISBN 3-540-09618-3

Blatter, Ch., Analysis II. Zweite verb. u. erweit. Aufl. mit Aufgaben. Heidelberger Taschenbücher Band 152. Berlin-Heidelberg-New York, Springer-Verlag 1979. IX, 196 S., 44 Abb., 1 Tab., DM 17,80, US $ 9.80 ISBN 3-540-09484-9

Budach, L. (Ed.), Fundamentals of Computation Theory – FCT'79. Proceedings of the Conference on Algebraic, Arithmetic, and Categorial Methods in Computation Theory held in Berlin/Wendisch-Rietz (GDR) September 17–21, 1979. Mathematical Research – Mathematische Forschung Band 2. Berlin, Akademie-Verlag 1979. 576 S., ca. 50 Abb., 2 Tab., 72,– M. Best.-Nr. 7628065 (2182/2)

Rao, M. M., Stochastic Processes and Integration. Alphen aan den Rijn-Germantown, Md., Sijthoff & Noordhoff 1979. XI, 456 S., Dfl. 110.00, US $ 55.00. ISBN 9028604383

Braun, M., Differentialgleichungen und ihre Anwendungen. Übers. a. d. Engl. von **T. Tremmel.** Hochschultext. Berlin-Heidelberg-New York, Springer-Verlag 1979. XII, 596 S., 65 Abb., 17 Tab. DM 48,--, US $ 26.40. ISBN 3-540-09391-5

Herman, G. T. (Ed.), Image Reconstruction from Projections. Implementation and Applications. Topics in Applied Physics 32. Berlin-Heidelberg-New York, Springer-Verlag 1979. XII, 284 S., 120 Abb., 10 Tab., DM 98,–, US $ 53.90. ISBN 3-540-09417-2

Devlin, K. J., Fundamentals of Contemporary Set Theory. Universitext. Berlin-Heidelberg-New York, Springer-Verlag 1979. VIII, 182 S., DM 19,–, US $ 10.50. ISBN 3-540-90441-7

Egardt, B., Stability of Adaptive Controllers. Lecture Notes in Control and Information Sciences 20. Berlin-Heidelberg-New York, Springer-Verlag 1979. V, 158 S., 16 Abb. 2 Tab., DM 21,50, US $ 11.90. ISBN 3-540-09646-9

Böhling, K. H. / Spies, P. P. (Hrsg.), GI – 9. Jahrestagung. Bonn, 1.–5. Oktober 1979. Informatik-Fachberichte 19. Berlin-Heidelberg-New York, Springer-Verlag 1979. XII, 690 S., 133 Abb., 25 Tab., DM 56,60, US $ 31.10. ISBN 3-540-09664-7

Künzi, H. P. / Krelle, W. / von Randow, R., Nichtlineare Programmierung. Unter Mitwirkung von **W. Oettli.** Zweite, neubearb. u. erw. Aufl. Hochschultext. Berlin-Heidelberg-New York, Springer-Verlag, 1979. XIV, 262 S., 36 Abb., DM 49,–, US $ 27.00. ISBN 3-540-09343-5

NACHRICHTEN

Zum siebzigsten Geburtstag von LOTHAR COLLATZ

Am 6. Juli 1980 vollendet LOTHAR COLLATZ sein siebzigstes Lebensjahr. Für den großen Kreis seiner Schüler, seiner Freunde, seiner Mitarbeiter und Mitstreiter ist dieser Tag ein willkommener Anlaß, ihm für den Reichtum an wissenschaftlichen Erkenntnissen, die sie ihm verdanken, für seine unermüdliche fachliche und persönliche Hilfsbereitschaft, für sein mutiges und leidenschaftliches Eintreten für die Belange der Angewandten Mathematik und nicht zuletzt für die Wärme, die seine Persönlichkeit ausstrahlt und die sich auch in seinem liebenswürdigen Humor ausdrückt, herzlichst zu danken und ihm für die Zukunft weiteres so erfolgreiches Schaffen und persönliches Wohlergehen zu wünschen.

LOTHAR COLLATZ wurde vor siebzig Jahren als Sohn eines Geodäten in Arnsberg (Westf.) geboren. Sein Studium absolvierte er von 1928 bis 1933 an den Universitäten Greifswald, Göttingen, München und Berlin. In Berlin legte er 1933 die wissenschaftliche Staatsprüfung für das Lehramt an höheren Schulen ab, 1935 promovierte er dort. In diese Zeit fallen seine ersten Veröffentlichungen, deren wichtigste bereits Fehlerabschätzungen für Differenzenverfahren und Differenzenverfahren mit höherer Approximation behandeln. Im Jahre 1938 habilitierte er sich an der Technischen Hochschule Karlsruhe, 1943 wurde er als ord. Professor an die Technische Hochschule Hannover, 1952 an die Universität Hamburg berufen. Dem hier von ihm gegründeten und durch ihn zu Weltruf gekommenen Institut für Angewandte Mathematik ist er über seine 1978 erfolgte Emeritierung hinaus bis zum heutigen Tage treu geblieben.

Der Jubilar ist in der ganzen Welt als einer der bedeutendsten Vertreter der Angewandten, insbesondere der Numerischen Mathematik bekannt, geachtet und beliebt. Vier Universitäten: die Universität von Sao Paolo (1956), die Technische Hochschule Wien (1966), die Universität Dundee (1974) und die Brunel University London (1977) haben ihm die Ehrendoktorwürde verliehen; 1962 hat ihn die Deutsche Akademie der Naturforscher Leopoldina zu Halle/S. zum ordentlichen Mitglied berufen, gegenwärtig ist er einer ihrer beiden Obmänner für Mathematik. Sein wissenschaftliches Werk umfaßt bis jetzt über 150 Veröffentlichungen, darunter mehrere grundlegende Bücher, von denen hier nur sein schon 1945 erschienenes über „Eigenwertprobleme und ihre numerische Behandlung" — das später in das Buch „Eigenwertaufgaben mit technischen Anwendungen" (1949, 1961) übergegangen ist, — seine „Numerische Behandlung von Differentialgleichungen" (1951) und sein Werk „Funktitionalanalysis und Numerische Mathematik" (1964) besonders hervorgehoben seien. Ungezählt sind seine Vorträge, die er überall in der Welt auf Tagungen, Konferenzen und Kolloquien gehalten hat und noch hält und in denen er bereitwillig seine reichen, immer wieder in Neuland vorstoßenden und immer fruchtbaren Ideen in didaktisch wohldurchdachter Form und unter Heranziehung einprägsamer konkreter Beispiele darbietet. (Sein Freundeskreis möchte seine „Jahresberichte" mit den liebenswerten eigenhändigen Federzeichnungen, in denen er seine in aller Welt gesammelten Reiseeindrücke festhält, nicht mehr missen.) Seine Vortragstätigkeit paßt zu seinem Wirken als Hochschullehrer, dessen Erfolge sich nicht zuletzt darin widerspiegeln, daß zahlreiche seiner Schüler heute als Professoren an Universitäten und Hochschulen sein Werk weiterführen. Hervorzuheben sind darüber hinaus die international in hohem Ansehen stehenden Kolloquien, die er — zusammen mit namhaften Fachkollegen — im Mathematischen Forschungsinstitut in Oberwolfach durchführt, und die Sorgfalt, mit der er sich um die Veröffentlichung der dort gehaltenen Vorträge kümmert.

Ein Versuch, die Verdienste von LOTHAR COLLATZ zu würdigen, kann nicht an seiner grundsätzlichen Auffassung von der Mathematik und an der Bedeutung, die er ihrer Anwendung auf die mannigfachsten Probleme in ursprünglich außermathematischen Objektbereichen beimißt, vorbeigehen. Wie für jeden Mathematiker, dem die Anwendungen der Mathematik am Herzen liegen, gibt es für ihn keine Trennungslinie zwischen einer „reinen" und einer „angewandten" Mathematik, eher eine Trennungslinie, die durch die Reihen der Mathematiker hindurchgeht und diejenigen, die sich auf das Abstrakte beschränken, von denen scheidet, die außerdem die Beziehungen zwischen dem Abstrakten und den in der realen Umwelt gegebenen Konkreten beachten. LOTHAR COLLATZ vertritt seine Auffassung nicht nur im geschriebenen Wort in seinen Werken — man braucht dazu nur die Vorworte zu seinen Büchern zu lesen und zur Kenntnis zu nehmen, an welche Mathematiker der Vergangenheit er sich anschließt, — sondern auch mit großer Energie und Leidenschaftlichkeit im gesprochenen Wort wie erst in jüngster Zeit in einem vielbeachteten Diskussionsbeitrag zum Thema „Idee und Erfahrung" auf der Jahrestagung der Leopoldina im Herbst 1977 in Halle/S., auf den auch Herr GLASHOFF Anfang vorigen Jahres in Hamburg während der aus Anlaß der Emeritierung von LOTHAR COLLATZ veranstalteten Tagung über Numerische Mathematik in seiner Würdigungsansprache hingewiesen hat. Es sei auch an die humorvoll-drastische Formulierung erinnert, die COLLATZ im Jahre 1965 als seinerzeitiger Vorsitzender der GAMM auf deren Wiener Jahrestagung seiner Auffassung gegeben hat und die hier, da sie schon fünfzehn Jahre zurückliegt und sich vornehmlich an die jüngere Generation wendet, auszugsweise wiederholt sei (s. ZAMM **45** (1965), T1):

„... die Einführung abstrakter Ideen und Methoden hat zu einer ungeheuren und früher nicht für möglich gehaltenen Erweiterung des Bereiches konkreter beherrschbarer Probleme geführt; der Zug zur Abstraktion und zur Verallgemeinerung ist nicht nur für die moderne Mathematik, sondern auch für alle modernen Naturwissenschaften kennzeichnend. Aber nun hat sich das merkwürdige Phänomen herausgestellt, daß die Beschäftigung mit dem Abstrakten die Gefahr in sich birgt, daß die Beherrschung des Konkreten, für die ja das Abstrakte geschaffen wurde, verloren zu gehen droht. ...

Ich möchte hier eine kleine ... Geschichte von dem Physiker und Nobelpreisträger PAULI erzählen: PAULI wohnte einmal einem sehr abstrakt gehaltenen Vortrag über Theoretische

Physik bei; in der Diskussion fragte jemand den Vortragenden, wie denn nun diese Theorie anzuwenden sei und was für Aussagen sich z. B. für einfache Atome ergäben. Der Vortragende antwortete: „Natürlich muß man das noch untersuchen, aber das sind Details, die man später noch einfügen kann." Da stand PAULI auf, zeichnete mit Kreide ein großes Rechteck an die Tafel und erklärte: „Das hier ist ein Gemälde von RAFFAEL, man hat nur noch die Details einzufügen." Und so möchte ich an alle Vortragenden die Bitte richten, daß sie die Details nicht vergessen und nicht nur am Rahmen arbeiten, wenn es auch oft bequemer ist. Wir brauchen in der angewandten Mathematik beides, Arbeit am Rahmen und Arbeit an den Details. Als sehr gesundes Maß für die Verallgemeinerungen möchte ich Ihnen den Ausspruch E. CARTANS nennen, es sei nicht die Absicht BOURBAKIS, Verallgemeinerungen nur um der Verallgemeinerung willen zu bilden, sondern „eine allgemeine Fassung hat nur dann Recht zu bestehen, wenn sie auf mehrere Fälle angewendet werden kann" (Hamburg, am 5. 6. 1957). ..."

Als langjähriger Herausgeber der ZAMM bin ich in meiner Zusammenarbeit mit dem Jubilar dieser seiner Auffassung immer wieder begegnet und habe ich ihr zu folgen versucht. Er gehört seit 1933 mit über dreißig Arbeiten zu den Autoren der ZAMM und seit 1955 zu den Mitwirkenden des Herausgebers. Mit meinem Kollegen GÜNTER SCHMIDT, dem jetzigen Herausgeber der ZAMM, bin ich darüber einig, wieviel wir dem Jubilar verdanken: nicht nur seine Buchbesprechungen, sondern vor allem seine Gutachten über eingereichte Manuskripte, auf die er viel — notwendig unsichtbar bleibende — Zeit und Mühe verwendet, sind in höchstem Grade sachlich, objektiv und zuverlässig und zwar streng, aber von einer wohlwollenden Strenge, die für den jeweiligen Autor und für den Herausgeber bei der Entscheidung, die er zu treffen hat, eine wesentliche Hilfe bedeutet. Auch dafür sei ihm an dieser Stelle aufrichtig gedankt.

Dresden Helmut Heinrich

www.ingramcontent.com/pod-product-compliance
Lightning Source LLC
LaVergne TN
LVHW081350011026
849161LV00013BB/3043